W9-ADZ-745

HIGH-PERFORMANCE LIQUID CHROMATOGRAPHY

Advances and Perspectives

Volume 3

CONTRIBUTORS TO THIS VOLUME

Phyllis R. Brown
William S. Hancock
Milton T. W. Hearn
Hubert A. Scoble
Lloyd R. Snyder
James T. Sparrow

HIGH-PERFORMANCE LIQUID CHROMATOGRAPHY

Advances and Perspectives

Volume 3

Edited by

Csaba Horváth

Department of Engineering and Applied Science
Yale University
New Haven, Connecticut

1983

ACADEMIC PRESS
A Subsidiary of Harcourt Brace Jovanovich, Publishers
New York London
Paris San Diego San Francisco São Paulo Sydney Tokyo Toronto

ACADEMIC PRESS, INC.
111 Fifth Avenue, New York, New York 10003

United Kingdom Edition published by
ACADEMIC PRESS, INC. (LONDON) LTD.
24/28 Oval Road, London NW1 7DX

ISBN 0-12-312203-1

ISSN 0270-8531
This publication is not a periodical and is not subject to copying under CONTU guidelines.

PRINTED IN THE UNITED STATES OF AMERICA

83 84 85 86 9 8 7 6 5 4 3 2 1

7596952

CONTENTS

CONTRIBUTORS

Numbers in parentheses indicate the pages on which the authors' contributions begin.

Phyllis R. Brown (1), Department of Chemistry, Pastore Chemical Laboratory, University of Rhode Island, Kingston, Rhode Island 02881

William S. Hancock (49), Department of Chemistry, Biochemistry, and Biophysics, Massey University, Palmerston North, New Zealand

Milton T. W. Hearn (87), St. Vincent's School of Medical Research, Melbourne, Victoria 3065, Australia

Hubert A. Scoble* (1), Department of Chemistry, Pastore Chemical Laboratory, University of Rhode Island, Kingston, Rhode Island 02881

Lloyd R. Snyder (157), 2281 William Court, Yorktown Heights, New York 10598

James T. Sparrow (49), Department of Medicine, Baylor College of Medicine, and The Methodist Hospital, Texas Medical Center, Houston, Texas 77030

*Present address: Department of Chemistry, Massachusetts Institute of Technology, Cambridge, Massachusetts 02139.

Every scientific advance is an advance in method.
M. S. TSWETT

CONSPECTUS

The analytical technique known in most languages as HPLC, the acronym for *high-performance liquid chromatography*, may very well represent the climax of the development started when the Italo-Russian botanist Michael S. Tswett coined the name chromatography and recognized the potential of his method for separating plant pigments almost 80 years ago. In order to distinguish HPLC from conventional column chromatography, the names high-pressure or high-speed liquid chromatography are also used occasionally. It is a separation method of unsurpassed versatility and a microanalytical tool *par excellence*. Like gas chromatography, HPLC is characterized by a linear elution mode and by the use of a sophisticated instrument, high-efficiency columns, and sensitive detectors.

Recent advances in instrumentation, column engineering, and theory have considerably broadened the field of application of HPLC, which now finds employment in virtually all branches of science and technology. Yet, we may have witnessed only the beginning of a long growth period in which HPLC will become the preeminent method of chemical analysis.

The goal of this serial publication is to provide up-to-date accounts of various topics in HPLC. The individual chapters will cover subjects of particular interest in this rapidly growing field. Throughout the successive volumes, the coverage of applications, instrumentation, and theory will be balanced, although the contents of some volumes may focus on one or the other of these subjects. As the field evolves and the horizon of HPLC expands, future volumes are expected to present full accounts of the advances in HPLC and to unfold the perspective required for exploiting its full potential.

Selection of topics and the level of treatment, at least in the early volumes, are planned to offer useful reading both for the novice and the seasoned chromatographer. Thus the contents will reflect not only the individuality of expression of the contributors, but also the diversity and broad scope characteristic of HPLC.

New Haven, Connecticut CSABA HORVÁTH

PREFACE

Since the commencement of this serial publication high-performance liquid chromatography (HPLC) has continued its meteoric growth, and HPLC is now safely entrenched as the premier analytical technique for mixtures of nonvolatile substances. During the past three years the acceptance of HPLC in the life sciences and the expansion of its scope to the rapid separation of biopolymers has been perhaps the most momentous event. The exploitation of the potential of reversed-phase chromatography (RPC) with hydrocarbonaceous bonded phases as a versatile, efficient, and convenient technique is particularly noteworthy in this regard. As it stands now, HPLC has become an indispensable tool in the armamentarium of life scientists and has found wide use on a quotidian basis.

The present Volume 3 bears an inscription of this development. The first article by Scoble and Brown gives an account of the analysis of nucleic acid fragments by RPC to serve the interest of a crescive audience engaged in analysis of nucleic acid constituents. The use of RPC for separation of proteins is the subject of the second contribution by Hancock and Sparrow. Although this application of HPLC is inchoate and still very much in flux, the significance of the subject and the broad interest associated with it fully warrant a treatment by noted experts in this field. In the ensuing article HPLC of peptides is reviewed by Hearn who examines the use of various techniques including bonded-phase, ion-exchange, and size-exclusion chromatography. There have been major advances in theory of liquid chromatography in many areas, and further progress on the practical side is likely to benefit from a firm theoretical footing. In the last article on mobile-phase effects in liquid–solid chromatography, Snyder presents a comprehensive view of the results of his fundamental studies on the physicochemical aspects of the retention process.

I would like to express my special thanks to the contributors to this volume and to appeal to the reader's indulgence for deferring the treatment of other timely and pertinent subjects to the following volumes of this publication.

Cs. H.

CONTENTS OF PREVIOUS VOLUMES

HIGH-PERFORMANCE LIQUID CHROMATOGRAPHY

Advances and Perspectives

Volume 3

REVERSED-PHASE CHROMATOGRAPHY OF NUCLEIC ACID FRAGMENTS

Hubert A. Scoble* and Phyllis R. Brown

Department of Chemistry
Pastore Chemical Laboratory
University of Rhode Island
Kingston, Rhode Island

* Present address: Department of Chemistry, Massachusetts Institute of Technology, Cambridge, Massachusetts 02139.

HIGH-PERFORMANCE LIQUID CHROMATOGRAPHY, Vol. 3

ISBN 0-12-312203-1

I. INTRODUCTION

A. Historical Background

The determination of nucleotide, nucleoside, and base concentrations is of prime importance in genetic, biomedical, and biochemical research. As our knowledge and interest in each of these areas has expanded, so has the demand for reliable analytical techniques for the rapid separation and determination of nano- and picomole quantities of free nucleotides, nucleosides, bases, and oligonucleotides.

The use of high-performance liquid chromatography (HPLC) as an analytical separation technique has had an explosive growth in the biochemical literature. The many modes of HPLC permit the rapid separation of widely varying classes of compounds. In addition, since compounds which are ionic, nonvolatile, or thermally labile can be analyzed by HPLC, derivatization prior to chromatographic separation is not usually needed.

The free nucleotides, nucleosides, and bases which are present in physiological fluids result from the catabolism of nucleic acids, enzyme-catalyzed degradation of bodily tissues, anabolic pathways such as the "de novo" or salvage pathways, or dietary intake.

In addition to major nucleosides and bases, modified nucleosides and bases have also been isolated from tRNA hydrolysates and in physiological fluids of man. Unlike the major nucleic acid components, the methylated or otherwise structurally altered purine and pyrimidine compounds are not recycled in the salvage pathways but are excreted. It has been suggested that the measurement of these modified compounds may provide an indicator of the rate of tRNA metabolism. Furthermore, the altered patterns of excretion for these compounds may be used as biomarkers for the detection of disease states and aberrations in metabolic pathways.

Early analyses of nucleic constituents by ion-exchange chromatography were developed by Cohn (C8) at Oak Ridge National Laboratory. Cohn's chromatographic system employed totally porous polystyrenedivinyl benzene ion-exchange resins with nominal diameters of 100 μm. Although these resins had high capacities, they had slow analysis times and limited efficiencies. Nonetheless, Cohn's pioneering work showed the potential of polymeric resins in chromatographic analyses for nucleotides and led to the development of other separation systems.

Uziel *et al.* (U5) used porous divinyl benzene cation-exchange resins to separate the four nucleoside hydrolysates of RNA. Burtis *et al.* (B34, B35) used a smaller particle cation-exchange material for this separation and also investigated the various parameters that affected the separations.

Due to inherent limitations of the totally porous polymeric resins, a new type of particle called the pellicular resins or superficially porous packings were developed and applied to nucleoside and base separations by Horváth and Lipsky (H27), Kirkland (K14), and others (B34, H18). In these packings, the ion-exchange resin was polymerized on a solid glass or silica support particle, usually 40–70 μm in diameter. The diffusion distance of the solute molecule in the stationary phase was thus minimized. These packing materials were successful in increasing the efficiency of HPLC columns but had two serious limitations; the ion-exchange capacity of the pellicular packings was extremely low and packings were unstable because there were no chemical bonds between the stationary phase and the solid support.

Although most nucleotides, nucleosides, and bases can be separated using either cation- (M21) or anion-exchange (A4, H20) chromatography with totally porous packings, the separations may require several hours. Floridi *et al.* (F4) reported on the simultaneous separation of nucleosides, bases, and nucleotide mono-, di-, and triphosphates using an Aminex A-14 resin. Utilizing a linear gradient elution of both ionic strength and pH, complete resolution of all three classes of compounds was achieved in approximately 3 h.

Singhal and Cohn (S26) used exclusion chromatography with anion-exchange columns for the separation of nucleosides and bases. Brown *et al.* (B30) optimized this technique and obtained a separation of the naturally occurring free nucleosides and bases in cell extracts. Although these separations were useful, the analysis time of 2 h and relatively low efficiencies of the columns limited their usefulness in routine analyses.

The need for high-efficiency packing materials in HPLC led to the development of microparticulate, chemically bonded packing materials. In these packings, the particles typically have diameters of 3–10 μm and the ligates are bonded to spherical or irregularly shaped, totally porous silica supports. Due to the smaller particle diameters and porosity of these supports, there is an increase in surface area which results in increased exchange capacities and higher column efficiencies.

Microparticulate ion-exchange packings have been used extensively in the analysis of nucleotides (B1, C7, H14, L9, M8, P4, R4) and to a lesser extent nucleosides (B1, B19, P4, S20). In addition, microparticulate affinity gels, exclusion gels, etc. have been used for the separation of these compounds (G7, G8, H28, L1, L2, S6, S9, S24–S27).

The recent popularity of HPLC as an analytical technique in biochemical and biomedical research can be attributed to the development and introduction of microparticulate reversed-phase packings in which the hydrocarbon chain (octadecyl-, ocyl-, or di-) moieties are chemically bonded to a silica base (K5). At the present time, it is estimated that approximately 80% of all HPLC separations are performed in the reversed-phase mode. Reversed-phase high-performance liquid chromatography (RPLC) has many advantages over other modes of HPLC:

(1) Ionic, nonionic, or ionizable compounds can be readily separated; oftentimes all three groups can be separated in one analysis.
(2) Chromatographic analysis times are rapid and after gradient elution column reequilibration time is short.
(3) Selectivity can be controlled either through the proper choice of packing material or by minor variations in the mobile phase.
(4) Reversed-phase columns are efficient and stable, and separations are reproducible.
(5) Through the use of selective equilibria such as ion pairing with ligands or metal ions, ion suppression, or ligand exchange, ionic or ionizable compounds can be selectively retained.
(6) RPLC can be used for determining various physiochemical properties such as hydrophobicity, dissociation constant, and complex formation constants

In this article we will discuss the use of reversed-phase chromatography in the analysis of nucleic acid fragments. For in-depth reviews of chromatographic theory and general applications, the reader is referred to several reviews which are available (B24, B31, G5, K8, M16, S17, S23).

B. Structure and Nomenclature of the Purines and Pyrimidines

1. Pyrimidines

The pyrimidine bases are derived from the parent compound pyrimidine. The basic structures and numbering systems are shown in Fig. 1. The 2-, 4-, and 6-positions of the pyrimidine ring all exhibit a marked π electron deficiency. The 1- and 3-positions of the ring nitrogen

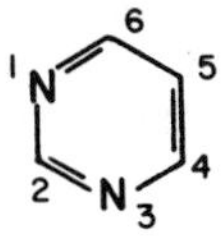

FIG. 1. Structure and systematic numbering of the pyrimidine base.

atoms reinforce this π electron deficiency, and as such, the 2-, 4-, and 6-positions are subject to nucleophilic attack. Due to inductive effects, the 5-position of the heterocyclic system is only slightly electron deficient. As a result, the 5-position is much more resistant to nucleophilic attack; however, if electrophilic substitution is to occur at all, the 5-position is most likely the site of attack.

With the introduction of electron-releasing substituents, as is the case with the biologically occurring pyrimidines, the π electron deficiency is counteracted in such a manner that the system approximates an aromatic ring. The 2-, 4-, and 6-positions are now deactivated toward nucleophilic attack and electrophilic substitution is facilitated.

If the 2-, 4-, and 6-position substituents are hydroxy, mercapto, or amino groups, the tautomers oxo, thio, or imino can form with the hydrogen being accommodated on a ring nitrogen. Experimental evidence indicates that the hydroxy group substituents prefer the oxo tautomeric form and mercapto groups the thio tautomeric form; however, the amino group normally exists as the free $—NH_2$. Tautomeric forms involving the 5-position are usually only seen when there is no ring nitrogen atom available. The effects of pyrimidine tautomeric forms on reversed-phase retention behavior have been studied by Brown and Grushka (B26) and will be discussed in later sections.

2. *Purines*

The purine bases are formed from the parent compound purine, which in turn is formed by the fusion of a pyrimidine and an imidazole ring. The parent purine structure and systematic numbering are shown in Fig. 2. Like the pyrimidine class of compounds, purine ring systems are π electron deficient; however, unlike pyrimidines, there is no position in the ring

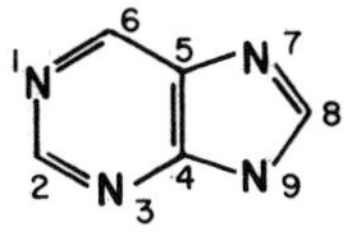

FIG. 2. Structure and systematic numbering of the purine base.

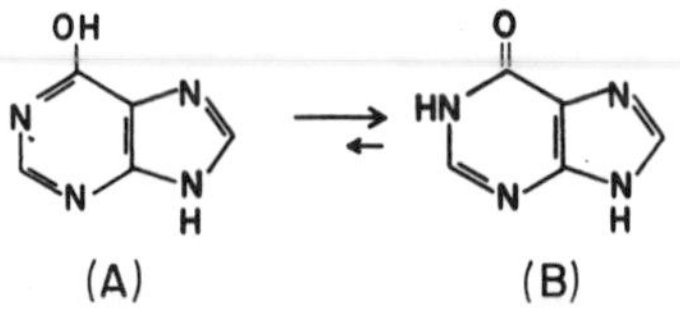

FIG. 3. (A) Lactim-imino and (B) lactam-amino tautomeric forms of the purine base hypoxanthine.

system equivalent to the 5-position of the pyrimidine ring. Thus, all ring carbon atoms which are electron deficient are activated toward nucleophilic attack and deactivated toward electrophilic attack.

As is the case for pyrimidines, purines also exist in oxo, thio, and amino tautomeric forms. The lactam-amino and lactim-imino forms of the purine base hypoxanthine are shown in Fig. 3.

3. Nucleosides and Nucleotides

Nucleosides consist of a heterocyclic aglygone attached via a glycosyl linkage to a carbohydrate moiety. Most biological nucleosides are *N*-glycosyl derivatives of the pentose sugars D-ribose or 2′-deoxy-D-ribose. Table I lists some of the common nucleosides. Nucleotides are phosphate esters of the nucleosides. Table II lists some of the common nucleotides.

In order to completely describe a nucleoside or nucleotide, the following information is needed:

(1) The chemical structure of the purine or pyrimidine base.
(2) The chemical structure and configuration of the carbohydrate moiety.
(3) The site of the attachment of the carbohydrate to the purine or pyrimidine base.

TABLE I

Major Nucleosides[a]

Common name	Base name	Common name	Base name
Adenosine	Adenine	Uridine	Uracil
Guanosine	Guanine	Cytidine	Cytosine
Inosine	Hypoxanthine	Thymidine	Thymine

[a] Carbohydrate moiety for thymidine is 2′-deoxy-D-ribofuranose; all others are D-ribofuranose.

TABLE II

Major Nucleotides

Ribonucleoside 5′-monophosphates:	
Adenosine 5′-phosphoric acid	(adenylic acid)
Guanosine 5′-phosphoric acid	(guanylic acid)
Cytidine 5′-phosphoric acid	(cytidylic acid)
Uridine 5′-phosphoric acid	(uridylic acid)
2′-Deoxyribonucleoside 5′-monophosphates:	
Deoxyadenosine 5′-phosphoric acid	(deoxyadenylic acid)
Deoxyguanosine 5′-phosphoric acid	(deoxyguanylic acid)
Deoxycytidine 5′-phosphoric acid	(deoxycytidylic acid)
Deoxythymidine 5′-phosphoric acid	(deoxythymidylic acid)

(4) The configuration of the glycosidic linkage.
(5) The number of phosphate groups and the site or sites of attachment.

The deoxynucleotides found in DNA include those of adenine, guanine, cytosine, and thymine; the ribonucleotides found in RNA contain adenine, guanine, cytosine, and uracil.

II. REVERSED-PHASE CHROMATOGRAPHY

A. Mechanisms of RPLC Retention

The exact mechanism(s) of solute retention in reversed-phase high-performance liquid chromatography (RPLC) is not presently well understood. The lack of a clear understanding of the mechanics of solute retention has led to a myriad of proposals, including the following: partition (K21, L6, S16); adsorption (C9, C10, H3, H15, H16, K13, L3, T2, U2); dispersive interaction (K2); solubility in the mobile phase (L7); solvophobic effects (H26, K6, M5); combined solvophobic and silanophilic interaction (B9, M12, N1); and a mechanism based upon compulsary absorption (B5).

From these investigations, it is apparent that no single retention mechanism is operative in RPLC; rather solute selectivity is based upon mixed-mode mechanisms. At present it appears that solvophobicity (hydrophobicity) is the primary mechanism for solute retention [Horváth *et al.* (B9, H23, H25, M12, M15, N1) and Karger *et al.* (K6, M5)].

The principle of solvophobicity as presented by Horváth *et al.* is based upon the tendency of the mobile phase to minimize the site of the cavity occupied by the solute molecules in the hydroorganic mobile phase. This can be viewed as a reversible association of the solute molecules with the hydrocarbonaceous stationary phase. The magnitude of the solvophobic effect for a given solute is due largely to the following four properties of the hydroorganic solvent system (H23):

(1) The ratio between the energy required for the production of a suitably shaped solvent cavity with a specific area, and the energy required to expand the planar surface of the solvent by this specified area.

(2) A constant, D, which is a function of the static dielectric constant, E, of the solvent expressed as

$$D = 2(E - 1)/(2E + 1)$$

(3) δ, the surface tension of the hydroorganic mobile phase.

(4) $\ln(RT/P_0V)$, which accounts for the entropy change associated with the change in "free volume."

Of these solvent properties, the dielectric constant and surface tension play important roles in governing solute retention.

The interaction of the solute with the mobile phase can bring about forces opposing those of the hydrophobic effect. In addition to van der Waals forces, which are dependent upon the size of the molecule involved, electrostatic interactions play a key role in solute retention. Solutes which have polar substituents can interact more strongly with the polar hydroorganic mobile phase, leading to a decrease in retention compared to similar compounds with no polar moiety. The ionization of a solute molecule under the appropriate mobile-phase conditions, results in an increase in electrostatic attraction between solute and eluent and ultimately to a decreased capacity for chromatographic retention.

Recently, Horváth and co-workers (B9, N1) introduced the concept of a dual binding mechanism to explain the atypical behavior of some solutes under reversed-phase conditions. In addition to solvophobic forces, it is possible for solutes to interact with the free surface silanols of the silica-based hydrocarbonaceous packing material. The term "silanophilic" interaction has been introduced to denote a reversible binding mechanism between solute molecules and silanol groups.

Horváth *et al.* (B9, N1) point out that, in addition to the hydrocarbonaceous ligates, silanol groups at the alkyl-silica stationary phase surface are accessible to solute molecules. The eluite can therefore independently bind in two different ways to the surface of a given stationary phase. In one mechanism, purely solvophobic interactions are responsible, whereas in the other, silanophilic interactions between solute and surface silanols are responsible.

This dual binding mechanism satisfactorily explains the anomalous retention of many solutes when viewed in purely solvophobic terms and will no doubt lead to the explanation of other solutes.

B. Structure–Retention Relationships of Purines and Pyrimidines in Reversed-Phase Systems

Although it has been difficult to rationalize the retention behavior of purine and pyrimidine compounds based upon their chemical structure and physiochemical characteristics, Brown and Grushka (B26) have recently formulated some rules for predicting retention behavior of these compounds. These rules are summaried below:

(1) Any substituent that causes charge formation decreases the capacity ratio k' of the compound.
(2) Any substituent that causes the compound to exist mainly as the lactam or amine tautomers decreases k'.
(3) The group type and position of the substituent on the ring affect the retention characteristics, with the order of the effect being OH $<$ H $<$ NH_2 $<$ NHR. A methyl substituent approximately doubles the k' value when compared with the parent compound.
(4) The addition of a ribosyl moiety to a base more than doubles k' when compared to the free base.
(5) In deoxyribonucleosides, the loss of the hydroxyl group from the 2′-position results in an increase of k' over the corresponding ribonucleoside.
(6) The addition of a linear phosphate group dramatically decreases k', but the addition of a cyclized phosphate group increases k' over that of the corresponding ribonucleoside.

A phenomenon of interest that is unique to aromatic compounds containing heteroatoms is that of association through vertical stacking in aqueous solutions (B12, B13, B21, S10, S28, S31, T5–T8). The phenomenon of base stacking of purine and pyrimidine compounds in reversed-phase systems has been investigated by Brown and Grushka (B26). Base stacking is unlike base pairing in RNA and DNA in which hydrogen bonding is the dominant force. The primary mechanism for base stacking is believed to be π electron overlap (B33) and as such, can be either heterogeneous, i.e., between a purine and a pyrimidine, or homogeneous, i.e., between the same type compound. It was found that the order of retention of nucleotides, nucleosides, and bases correlated to the order of stacking as was determined by calculations involving the free energy (ΔG) and equilibrium constants of stacking (T5–T8). As yet, it has not been determined whether stacking affects retention or whether the same factors that control retention also control stacking. However, at the concentrations of nucleosides and bases in physiological fluids, it appears that stacking cannot affect net retention but that the solvophobic factors involved in retention are operative in stacking.

C. Effects of Mobile Phase on Purine and Pyrimidine Retention Characteristics

The mobile-phase characteristics affecting the reversed-phase high-performance liquid chromatographic retention behavior of purine- and pyrimidine-based compounds have been investigated by Gehrke *et al.* (G3), Hartwick *et al.* (H6, H13), and most recently by Zakaria and Brown (Z1). In the investigation by Zakaria and Brown (Z1) the roles of the mobile-phase parameters pH, ionic strength, and percent organic modifier have been examined in reversed-phase chromatographic systems. The results of this and similar studies corroborate the earlier investigations by Brown and Grushka (B26), Horváth *et al.* (B9, H21–H23, H25, M13, M14, N1), and others (K6).

1. Mobile-Phase pH Effects

The retention behaviors of purine and pyrimidine bases as a function of pH (G3, H6, H13, Z1) are in agreement with the observations of Brown and Grushka (B26) that compounds existing mainly as the aromatic tautomer will have a greater tendency for vertical association than will those compounds which exist in the nonaromatic configuration. As the pH of the mobile phase is varied and the ratio of neutral to ionized species increases, the retention of the compound will also increase.

With the introduction of the electron-withdrawing ribose moiety, nucleosides have a greater tendency for self-association through stacking than their corresponding purine or pyrimidine base. This effect will be reflected in an increase in retention behavior of nucleosides over that of corresponding bases at any mobile-phase pH.

Zakaria and Brown (Z1) have found that, whereas nucleoside and base retention mechanisms can be adequately explained in terms of solvophobic considerations, nucleotide retention behavior can best be explained in terms of a mixed-mode mechanism. In an acidic mobile phase, it has been observed that ribonucleotides elute in order of increasing negative charge. This elution pattern is atypical for the reversed-phase mode and can perhaps be explained in terms of silanophilic (B9, N1) interactions of nucleotide phosphate moieties with free surface silanols of the stationary phase.

2. Organic Modifier Mobile-Phase Effects

Whereas nucleic acid constituents tend to vertically associate in aqueous solutions, the introduction of organic solvents tends to disrupt this associative process, leading to a ''destacking'' of the compounds.

The retention behavior of nucleic acid constituents in the presence of increasing concentrations of an organic modifier is typical of that expected based upon solvophobic mechanisms; with increasing concentrations of organic modifier, the capacity for solute retention is decreased. As expected, the use of a stronger eluent under reversed-phase conditions (e.g., acetonitrile vis-à-vis methanol) results in a subsequent decrease in the capacity ratios of all compounds in the chromatographic analysis.

In terms of the effects of organic modifier on nucleic acid constituent retention behavior, the nucleotides again exhibit anomalous retention characteristics. At low pH values and increasing concentrations of organic modifier, the order of retention is ribonucleotide monophosphates < ribonucleotide diphosphates < ribonucleotide triphosphates. This contradicts the solvophobic mechanistic retention view since the triphosphates have three ionized phosphates at low pH and are eluting after the monophosphates which have one ionized phosphate group at low pH. Again, silanophilic considerations may help in explaining this anomalous behavior. Since the use of a stronger eluent causes a subsequent decrease in retention capacity for the nucleotides, solvophobic considerations may be partially responsible for nucleotide retention. Clearly, it seems likely that a mixed-mode retention mechanism is in effect.

3. Effects of Mobile-Phase Ionic Strength

As reported by Zakaria and Brown (Z1), changing the mobile-phase ionic strength has little effect on the retention behavior of nucleosides and bases. The effect of mobile-phase ionic strength is much more pronounced in the case of ribonucleotides, where the compounds are ionized under the mobile-phase conditions.

With increasing ionic strength the ribonucleotide triphosphates, which have the largest negative charge, are subject to the greatest decrease in retention. Also, ribonucleotides with protonated bases exhibit a smaller change in k' than those with neutral bases. A possible explanation for this phenomenon would be a net decrease in the total ribonucleotide charge due to interaction between the protonated base and the negatively charged phosphate moiety.

D. Ion-Pairing Techniques

The use of ion-pairing techniques with reversed-phase liquid chromatography has extended the usefulness of RPLC to include simultaneous analyses involving ionized and nonionized compounds (B6, K3, M9).

The chromatographic use of ion-pairing techniques, in which amphiphilic ions are added to the mobile phase, resulted from the early work of Schill *et al.* (S4, S5) with further application by others (B18, E2, E3, K7). In addition, significant contributions to the theory and chromatographic application of ion pairing have resulted from the work of Haney *et al.* (S32, W4), Knox *et al.* (K16–K21), Horváth *et al.* (H24, M10, M11), and Bidlingmeyer *et al.* (B7, B8). Although the exact mechanism of reversed-phase ion pairing has been the subject of controversy, three general mechanistic views have been postulated.

The first view suggest the formation of ion pairs in the aqueous mobile phase prior to adsorption onto the chemically bonded stationary phase (H24, H26, W4). The capacity for solute retention is determined by the degree of hydrophobicity of the ion pair. Thus, the use of longer alkyl chain pairing agents imparts a higher degree of hydrophobicity, thereby increasing the retention of the solute molecules.

The second approach suggests that the ion-pair phenomena is based upon an ion-exchange mechanism (D4, H19, K15, K16, K19, K20, K25, K26, S14, T4, V2). In this theory, the unpaired lipophilic ions are adsorbed onto the nonpolar stationary phase in such a manner as to create a dynamically coated ion exchanger. Again, the longer the alkyl chain length on the ion-pairing reagent, the longer will be the retention of the solute.

The third mechanistic approach has been proposed by Bidlingmeyer *et al.* and Deming *et al.* (B6–B8, K24). In the ion-interaction approach ion pairs *do not* form in the mobile phase; rather it is assumed that there is a dynamic equilibrium of lipophilic ions. This results in the formation of an electrical double layer on the hydrocarbonaceous stationary phase. Retention is based, therefore, upon an electrostatic attraction due to surface charge density of the ion-pairing ions and from a "sorption" effect onto the nonpolar stationary phase.

At present, the phenomena occurring in ion-pair chromatography cannot be adequately explained either by a simple ion-pairing mechanism or the dynamic ion-exchange mechanism.

It is apparent that debate concerning the reversed-phase ion-pair retention mechanism will continue because of the paucity of pertinent data concerning ion-pair stability constants in hydroorganic solutions.

Reversed-Phase Ion Pairing of Purine- and Pyrimidine-Based Compounds

The specific use of reversed-phase ion-pairing techniques for the simultaneous analysis of purine and pyrimidine bases, nucleosides and, nucleotides has found limited chromatographic application, although it is a technique that warrants further investigation.

Hoffman and Liao (H19) first investigated the use of tetra-*n*-butylammonium hydrogen sulfate as an ion-pairing reagent for the separation of adenine, guanine, cytidine, and uracil ribonucleotides using a modified octadecylsilane (ODS) column. Although complete resolution of all components was not achieved, this work illustrated the utility and potential of the ion-pairing technique for these separations (Fig. 4).

Erhlich and Erhlich (E1) have reported on the separation of uracil, 5-methylcytosine, cytosine, guanine, and thymine using reversed-phase ion pairing (Fig. 5) and also, on the separation of adenine and N^6-methyladenine from the five DNA bases listed above. Although the separation was reported using only reference compounds, it has potential for the analysis of DNA hydrolysates.

Brown *et al.* (B23) separated uric acid, hypoxanthine, xanthine, and allopurinol by RPLC using heptane sulfonic acid as the ion-pairing agent. The chromatographic analysis is achieved in under 6 min with a reported minimum detection limit of 5 ng.

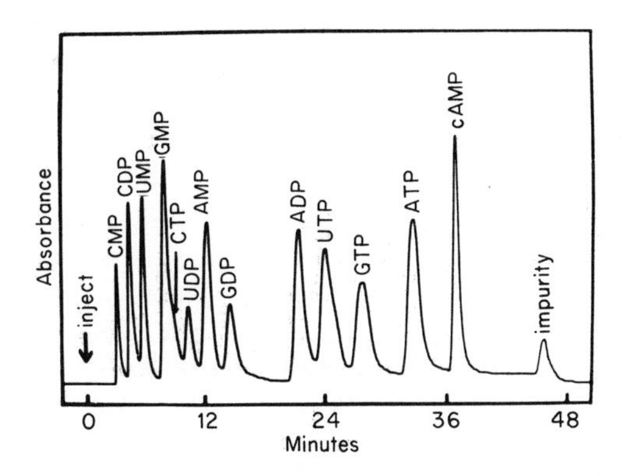

FIG. 4. Gradient elution separation of ribonucleotides. Mobile phase: (A) 0.025 *M* tetra-*n*-butylammonium hydrogen sulfate, 0.050 *M* KH_2PO_4, 0.080 *M* NH_4Cl buffered at pH 3.90; (B) 0.025 tetra-*n*-butylammonium hydrogen sulfate, 0.10 *M* KH_2PO_4, 0.20 *M* NH_4Cl buffered at pH 3.4, 30% methanol. Operating conditions: 40-min gradient program (concave #8) at 1 ml/min. Reprinted with permission from Hoffman and Liao (H19). Copyright by the American Chemical Society.

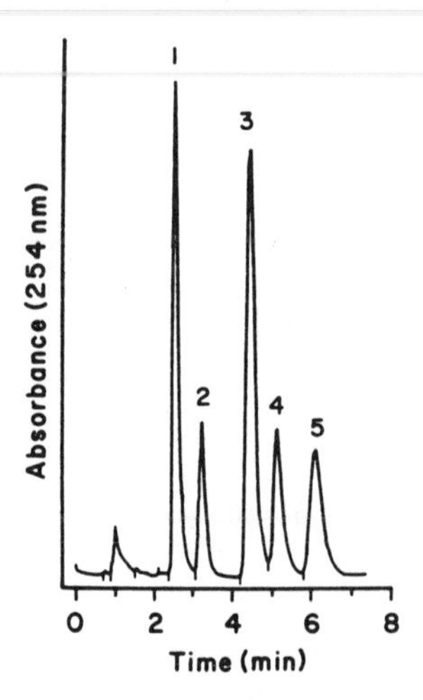

FIG. 5. Resolution of approximately 200 pmol each of uracil (1), 5-methylcytosine (2), guanine (3), cytosine (4), and thymine (5). The bases were chromatographed on an RP-18 column with 5 m*M* heptane sulfonate in 2.5 m*M* potassium phosphate, pH 5.6, at a flow rate of 2 ml/min. Reprinted with permission from Ehrlich and Ehrlich (E3). Copyright by Preston Publications, Inc., Niles, Illinois.

Juengling and Kammermeier (J2) reported the separation of purine nucleotides from rat heart extract using tetra-*n*-butylammonium phosphate on a C-18 column. Although the chromatographic separation has poor resolution, they emphasize the need for a chromatographic technique for the simultaneous analysis of ionic and nonionic compounds that retains the flexibility of the reversed-phase mode.

Walseth *et al.* (W3) separated 5′-ribonucleoside monophosphates using ion-pair chromatography. The chromatographic system has been used to analyze tissue nucleotides, to purify nucleoside monophosphates and cyclic nucleotides, and to monitor enzymatic reactions involving cyclic nucleotide phosphodiesterase, 5′-nucleotidase, and adenylate cyclase.

Purine drugs and antimetabolites have also been separated by ion-pairing techniques as reported by Day *et al.* (D3) and Voelter *et al.* (V4).

The use of metal ions to improve chromatographic selectivity has been investigated by Grushka *et al.* (C4, C5, G6, G8), Karger *et al.* (C11, L5), and Hare and Gil-Av (H4). Grushka *et al.* (C4) reported on the isocratic separation of nucleotides using a dithiocarbamate–Co(III) complexed column and Mg(II) in the mobile phase. More recently, Grushka and Chow (G8) reported on the separation of nucleosides and nucleotides using a dithiocarbamate column and Mg(II) in the mobile phase (Figs. 6

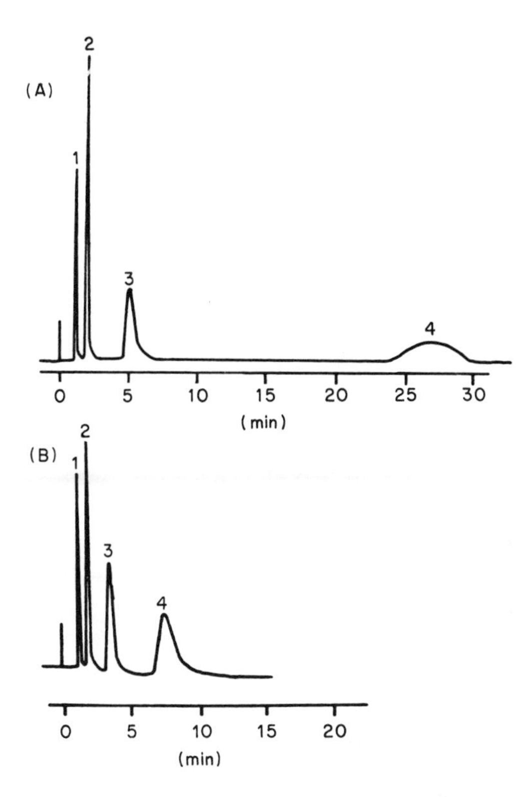

FIG. 6. Separation of some nucleosides and nucleotides. (A) Mobile phase: 0.137 *M* KH_2PO_4 with no Mg(II), pH 6.0, flow rate 1.5 ml/min. (B) Mobile phase: 0.137 *M* KH_2PO_4 with 8.1×10^{-4} *M* $MgSO_4 \cdot 7\,H_2O$, pH 6.0, flow rate 1.5 ml/min. Peaks: 1, uridine; 2, UMP; 3, UDP; 4, UTP. Reprinted with permission from Grushka and Chow (G8). Copyright by Elsevier Scientific Publishing Company, Amsterdam.

and 7), and Hubert and Porath (H29) reported on the use of copper chelate affinity chromatography for the separation of nucleotides.

III. SAMPLE PREPARATION

A. Introduction

The routine analysis of picomole quantities of nucleic acid fragments requires an increased compatibility between prechromatographic sample preparation procedures and sensitive liquid chromatographic detection

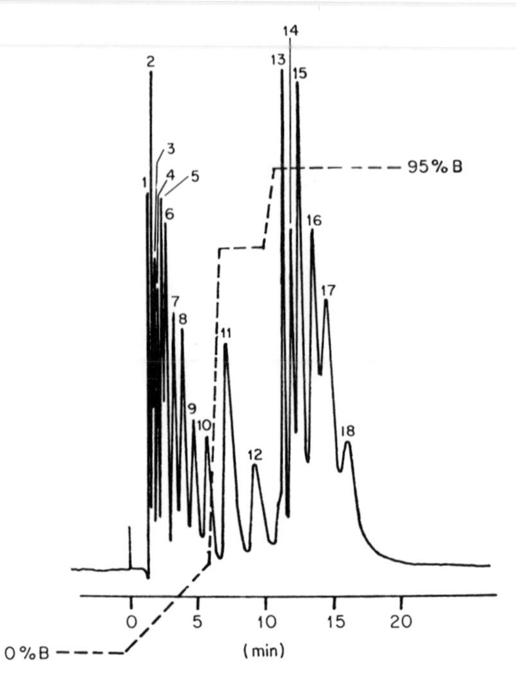

FIG. 7. Separation of some nucleosides and nucleotides by dithiocarbamate column using gradient elution. Mobile phase: (A) 0.137 *M* KH_2PO_4, 2.0 × 10^{-3} Mg(II) at pH 5.5. Mobile phase: (B) 0.137 *M* KH_2PO_4, 2.0 × 10^{-3} Mg(II) at pH 7.0. Gradient profile is shown by the dashed line. Peaks: 1, uridine; 2, thymidine; 3, guanosine; 4, adenosine; 5, deoxyadenosine; 6, UMP; 7, CMP; 8, dCMP; 9, GMP; 10, AMP; 11, XMP; 12, UDP; 13, CDP; 14, UTP; 15, CTP; 16, ADP; 17, XDP; 18, XTP. Reprinted with permission from Grushka and Chow (G8). Copyright by Elsevier Scientific Publishing Company, Amsterdam.

systems. In the analysis of biological samples the primary requirement is the removal of proteinaceous material which can become irreversibly bound to the analytical column. In addition to offering high analyte recovery, the sample preparation procedure should involve a minimum number of steps and minimal handling time and should not introduce interferents into the chromatographic analysis.

B. Boronate Affinity Gels

Immobilized boronic acid-substituted polymers have been used to improve the sensitivity and selectivity in RPLC analysis of ribonucleosides and ribonucleotides (D1, D2, G3, G4, K9, K38, L8, M19, U4). Boronate forms an anionic complex with vicinal 2′,3′-hydroxyls of unsubstituted ribonucleosides. The formation of the cyclic boronate esters occurs in alkaline conditions and is dependent on pH, ionic strength, and temperature. The structure of the boronate-derivatized polymer and the formation of the cis-diol boronate complex are shown in Fig. 8.

The use of group-specific ligands for the class separation of nucleosides has found widespread use in the analysis of plasma and urine from patients with various diseases (D1, D2, G3, G4, K38, P2, U4).

Gehrke *et al.* (G4) reported that nucleobases, deoxyribonucleosides, deoxyribonucleotides, and UV-absorbing pigments are all removed through the prechromatographic use of phenylboronate gels. They have further reported that major ribonucleotide monophosphates, although retained by the boronate gel, did not interfere with the liquid chromatographic analysis of the ribonucleosides (G4). Figure 9 illustrates the stepwise gradient elution of urinary ribonucleosides in which the sample pretreatment involved phenylboronate gels. Table III summarizes recovery studies for some ribonucleosides using boronate gel isolation prior to chromatographic analysis (G3).

Glad *et al.* (G7) reported on the use of a microparticulate boronic acid-substituted liquid chromatographic column in which ribonucleosides and ribonucleotides can be separated. Contrary to what Gehrke *et al.* (G4) reported, Glad *et al.* (G7) found that deoxynucleosides are retained even though they do not contain 2′,3′-cis-hydroxyl groups. This behavior is rationalized in terms of immobilized phenyl ring–nucleobase interaction in addition to the possibility of hydrogen bonding.

The use of group-specific sample preparation procedures has greatly improved the selectivity for the analysis of ribonucleosides in physiological fluids. In addition to reducing the demands made upon the chromatographic detection system, the technique is extremely well suited for the removal of potential interferents.

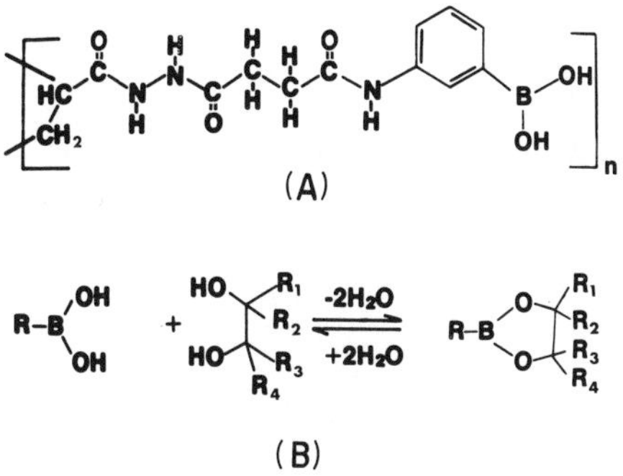

(A)

$$R{-}B(OH)_2 + HO{-}CR_1R_2{-}CR_3R_4{-}OH \underset{+2H_2O}{\overset{-2H_2O}{\rightleftharpoons}} R{-}B\langle O_2CR_1R_2CR_3R_4\rangle$$

(B)

FIG. 8. Structure of the boronate derivatized polymer (A) and the formation of the cis-diol boronate complex (B).

TABLE III

Recovery of Nucleosides Added to Pooled Control Urine Using Phenylboronate Gel[a,b]

Nucleoside	Concentration (nmol/ml)			Average recovery (%)
	Urine + spike	Urine	Spike recovered	
1-Methyladenosine	23.30	17.38	5.92	92
7-Methylguanosine	10.05	5.69[c]	4.36	88
Guanosine	13.47	8.91[c]	4.56	92
1-Methylinosine	15.46	10.48	4.98	99
1-Methylguanosine	10.66	5.64	5.02	101
N^2-Methylguanosine	10.69	5.46	5.23	100
Adenosine	7.38	2.55	4.83	98
N^2,N^2-Dimethylguanosine	15.72	11.28	4.44	100

[a] Reprinted with permission from Gehrke *et al.* (G4). Copyright by Elsevier Scientific Publishing Company, Amsterdam.

[b] Each value is an average of four runs.

[c] An unknown peak eluted with 7-methylguanosine and guanosine were integrated together.

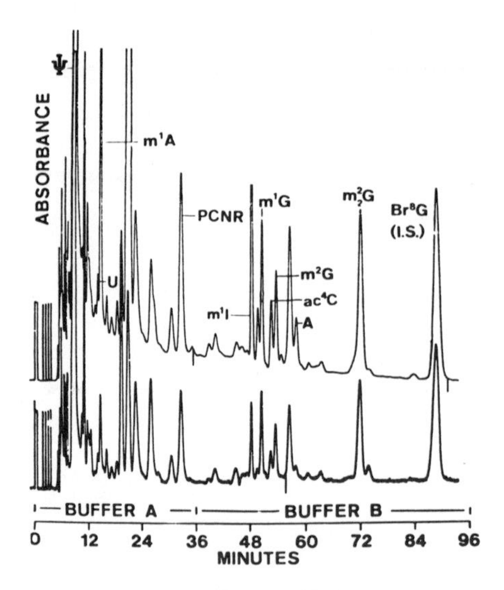

FIG. 9. Step-gradient reversed-phase HPLC separation of nucleosides in urine. Sample: 25 μl pooled ovarian cancer patient urine; column: μ-Bondapak C18, 600 × 4 mm; buffer, 0.01 *M* $NH_4H_2PO_4$: (A) pH 5.3, with 2.5% methanol, (B) pH 5.1, with 8.0% methanol; flow rate: 1.0 ml/min; temperature: 35°C; detection: 254 nm, 0.02 AUFS (upper trace); 280 nm, 0.01 AUFS (lower trace). Reprinted with permission from Gehrke *et al.* (G3). Copyright by Elsevier Scientific Publishing Company, Amsterdam.

C. Strong Acid Extraction

Trichloroacetic acid (TCA) and perchloric acid (PCA) has traditionally been used for the precipitation of proteinaceous materials from biological fluids. These methods of sample preparation have been used extensively for the analysis of nucleotides and nucleosides in physiological fluids (B1, C3, C7, D6, E5, H10, K9, K12, K27, K28, M3, M4, N3, R3–R5, S9, T1, V3, W5), and also for the analysis of tissue nucleotides (L9, R3).

To prevent possible nucleotide degradation when using TCA, Khym (K12) suggested the use of amine/Freon solution to back-extract excessive acid. Khym's procedure has been modified by Van Haverbeke and Brown (V3), and has been applied by others to the study of nucleotides (C3, R3, V3) and nucleosides (H10). Similar to the results reported by Chen *et al.* (C3), Riss and co-workers (R3) reported that the recovery of the adenosine-, cytidine-, guanosine-, and uridine-5′-nucleotides range from 81 to 99%, with an average recovery of 91% (Table IV).

Hartwick *et al.* (H10) found the TCA extraction procedure less suitable for the analysis of nucleosides and bases in serum, with analytical recoveries ranging from 54 to 68%. It is believed that a 15% loss in recovery could be attributed to nucleoside and base solubility in tri-*n*-octyl amine/Freon extraction medium.

The perchloric acid extraction procedure described by Kraus and Reinboth (K27) involves a perchloric acid extraction followed by neutralization with a potassium hydroxide solution. Riss *et al.* (R3) reported recoveries ranging from 68 to 87% for the adenosine-, cytidine-, guanosine-, and uridine-5′-nucleotides, with a mean recovery of 78%. Other investigators have used the PCA extraction procedure in the analysis of nucleotides

TABLE IV

Recovery of Nucleotide Standards Using Alamine/Freon Extraction Procedure[a]

Nucleotide	Alamine/ Freon extraction	Nucleotide	Alamine/ Freon extraction	Nucleotide	Alamine/ Freon extraction
Ribonucleoside monophosphates:		Ribonucleoside diphosphates:		Ribonucleoside triphosphates:	
AMP	89.3 ± 1.9	ADP	96.3 ± 0.4	ATP	91.1 ± 1.7
CMP	99.3 ± 1.3	CDP	88.9 ± 2.2	CTP	90.6 ± 1.7
GMP	81.0 ± 5.9	GDP	90.5 ± 3.5	GTP	89.8 ± 2.5
UMP	88.8 ± 1.0	UDP	94.0 ± 0.7	UTP	89.8 ± 0.7

[a] Reprinted from Riss *et al.* (R3) by courtesy of Marcel Dekker, Inc., New York.

(B1, E5, H7, L9, N3, R4, S9) and some nucleosides (B1, K9, N3, R5, T1, W5).

D. Ammonium Sulfate, Reversed-Phase Cartridges, Ultrafiltration

Hartwick *et al.* (H10) studied the use of high-ionic-strength solutions for the precipitation of serum proteins. In this technique a saturated solution of ammonium sulfate was added to serum and the supernatant was analyzed by RPLC after centrifugation. The recoveries for serum nucleosides and xanthine alkaloids were good, ranging from 93 to 100%. In addition, the serum profile showed no interference from this extraction technique.

The use of reversed-phase cartridges for serum sample preparation was also studied by Hartwick *et al.* (H10) and excellent recoveries were reported for non-protein-bound compounds.

The results of several methods for sample deproteinization prior to chromatographic analysis are shown in Table V.

The technique of ultrafiltration for serum sample preparation has been studied by several investigators (H7, H10, K9, S19, V4, V5). In this technique the serum sample is passed through an exclusion membrane which exhibits a 95% retention for compounds with MW > 25,000. Thus, low-molecular-weight constituents pass through the membrane, whereas proteins or protein-bound constituents are retained. Hartwick *et al.* (H10) reported excellent recoveries for xanthosine, inosine, guanosine, theobromine, theophylline, and caffeine. Tryptophan showed poor quantitative recovery in accordance with the findings of previous investigators (K33, O1).

E. Summary

Sample preparation is a crucial step in the analysis of biological samples. The methods of sample preparation that were presented here are only a few of the more common procedures that have appeared in the scientific literature.

The use of trichloroacetic acid followed by amine/Freon extraction appears to be the method of choice in the analysis of erythrocyte or tissue nucleotides. In addition to a high recovery for many ribonucleotides, the method is most effective at removing proteinaceous material.

For the analysis of ribonucleosides the use of phenylboronate affinity gels is the most selective. If, however, in addition to ribonucleosides the investigator wishes to analyze for purine or pyrimidine bases, other methods of sample preparation must be investigated.

TABLE V

Recoveries of Compounds Added to Pooled Human Serum[a]

Compound	TCA at		$(NH_4)_2SO_4$	Ultrafiltration at		Precolumn
	6%	12%		pH 7.8	pH 5.1	
Xanthosine	58.8 ± 6.2	66.4 ± 1.0	97.4 ± 7.7	99.3 ± 2.5	74.8 ± 4.2	—[b]
Inosine	59.4 ± 7.7	65.9 ± 2.7	95.6 ± 7.6	98.9 ± 2.2	97.7 ± 1.7	101.0 ± 2.8
Guanosine	54.0 ± 4.7	67.6 ± 0.3	84.8 ± 9.6	73.6 ± 3.9	85.9 ± 6.4	92.0 ± 5.3
Tryptophan	—	—	102.0 ± 12.6	12.1 ± 2.6	96.2 ± 3.7	43.3 ± 2.1
Theophylline	—	—	82.7 ± 5.1	40.7 ± 5.0	85.5 ± 4.7	91.2 ± 4.8
Theobromine	—	—	72.1 ± 7.5	81.3 ± 3.2	83.2 ± 4.3	90.2 ± 5.4
Caffeine	—	—	88.9 ± 6.7	83.0 ± 9.1	87.7 ± 4.5	92.8 ± 5.7

[a] Reprinted from Hartwick *et al.* (H10) by courtesy of Marcel Dekker, Inc., New York.
[b] Recovery variable, pH dependent.

In the analysis of nucleosides, bases, and other low-molecular-weight compounds, the use of ultrafiltration is the most reliable technique. In addition to minimizing the number of steps, the use of ultrafiltration does not require dilution of the original sample.

It should be noted that with the use of a column guard to protect the analytical column from irreversible adsorption of proteins, it is sometimes possible to inject serum, urine, or other physiological fluids directly into the chromatograph without prior sample processing, especially if very small samples are used.

IV. PEAK IDENTIFICATION

A. Introduction

Improvements in column technology, detector sensitivity and the development of new detection systems, have made possible the routine separation of picomole quantities of nucleic acid components in complex physiological matrices. The very sensitivity of most LC systems, however, which is invaluable in the analysis of biological samples, is often the limiting factor because of inadequate or ambiguous identification methods. Although tremendous advances have been made in the on-line combination of HPLC with spectroscopic techniques [e.g., mass spectrometry, Fourier transform infrared (FT/IR), nuclear magnetic resonance], their application has not become routine in most biochemical and biomedical laboratories.

Faced with the practical problems of determining the chemical nature of chromatographic peaks in biological matrices, multiple identification techniques are used in order to unambiguously identify eluting compounds. These techniques are summarized in Table VI.

B. Retention Behavior and Co-Chromatography with Reference Compounds

The retention behavior of the eluite alone cannot be used for the unambiguous identification of the compound under study. This is especially important in the analysis of physiological fluids where several different compounds in the matrix may have similar retention characteristics. Thus, a comparison of retention times for the eluting compounds with those of reference compounds that have been previously chromatographed under identical conditions, can only be used for tentative identification. For this identification procedure, knowledge of the type of compounds and their relative levels in the sample is usually necessary.

The reversed-phase chromatographic analysis of a control serum sample is shown in Fig. 10. A majority of the chromatographic peaks in this

TABLE VI

Methods of Peak Identification in HPLC

1. Peak retention data
2. Co-chromatography with reference standards
3. Comparison of compound retention behavior with reference standards under various chromatographic conditions
4. Enzymatic peak shift
5. Chemical derivatization
6. Spectral characterization
    a. Fixed wavelength
    b. Absorbance ratios
    c. UV–visible scans
    d. Fluorescence detection
    e. Fluorescence excitation and emission spectra
7. Miscellaneous techniques
    a. Characterization of collected fractions using spectroscopic techniques
    b. Electrochemical detection
    c. On-line mass spectrometry and IR spectroscopy
    d. Laser-induced fluorescence
    e. Photodiode array detectors
    f. Radioactivity detectors

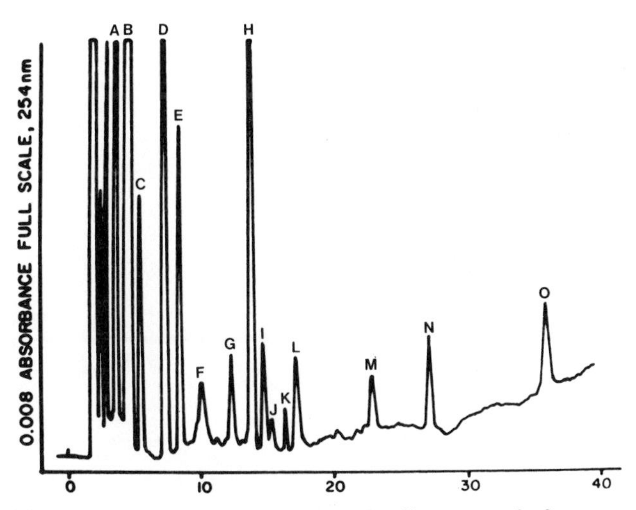

FIG. 10. Chromatogram of a control serum ultrafiltrate sample from a normal donor using UV detection at 254 nm. Injection volume: 80 μl. Column: reversed-phase, 10-μm particle diameter. Eluents: (A) 0.02 M KH_2PO_4, pH 5.6; (B) 60% methanol–water. Gradient: linear, 0–100% of B in 87 min, slope 0.60% methanol/min. Flow rate: 1.5 ml/min. Reprinted with permission from Hartwick *et al.* (H7). Copyright by Elsevier Scientific Publishing Company, Amsterdam.

sample have been tentatively identified by comparing their retention times with those of previously chromatographed reference compounds. Figure 11 represents the same serum sample to which standard reference compounds have been added. The co-injected chromatogram should be carefully examined for the appearance of shoulders or the doubling of peaks, which would be indicative of compounds with similar yet not identical retention times. Again, it should be emphasized that co-chromatography cannot ensure the unambiguous identification of compounds.

Further evidence for the characterization of compounds can be obtained by comparing the retention times of unknown compounds with those of the reference standards under different chromatographic conditions. This is not practical with multicomponent analyses where the chromatographic operating conditions have already been optimized for resolution, selectivity, and time of analysis.

C. Spectroscopic Data

1. Absorbance Ratios

The use of on-line UV–visible spectroscopic techniques has provided chromatographers with a sensitive and selective method for the charac-

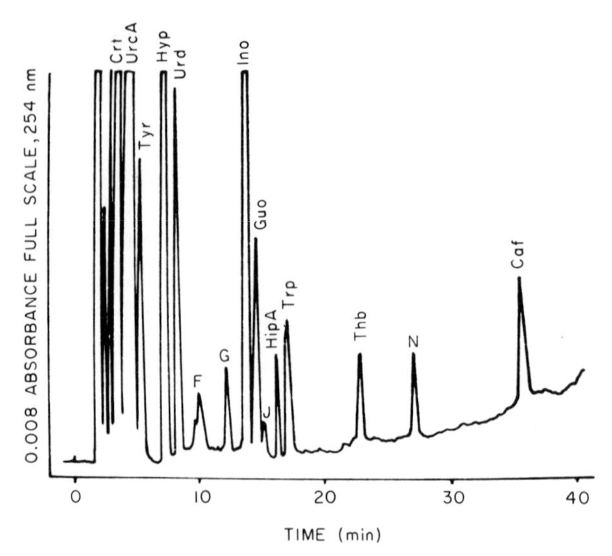

FIG. 11. Control sample co-injected with a solution containing creatinine (Crt), uric acid (UrcA), tyrosine (Tyr), hypoxanthine (Hyp), uridine (Urd), inosine (Ino), guanosine (Guo), hippuric acid (HipA), tryptophan (Trp), theobromine (Thb), and caffeine (Caf). Conditions same as in Fig. 10. Reprinted with permission from Hartwick *et al.* (H7). Copyright by Elsevier Scientific Publishing Company, Amsterdam.

terization of eluting compounds. UV–visible detectors have a wide linear dynamic range coupled with their high signal-to-noise ratio, which has led to their increased use in the analysis of biological constituents since many groups of biologically active compounds have chromophores which absorb in the UV or visible region.

The use of absorbance ratios for the monitoring of chromatographic effluents has been an increasingly popular tool for the spectral characterization of chromatographic peaks (B2, B20, B28, B29, C7, D1, E6, F1, G3, H7, H8, H12, H13, K9, K31–K33, K37, L4, L8, M3, M17, P3, T1, W5, Y3, Y4). Through the use of dual-wavelength of sequential detection systems, which are set to monitor different wavelengths, it is possible to calculate absorbance ratios for eluting compounds. Absorbance ratios are particularly important in distinguishing between compounds with similar retention characteristics. As is often the case, compounds with similar retention behavior exhibit distinctive absorbance ratios by which the compounds can be differentiated.

Criteria for selecting wavelengths for the measurement of absorbance ratios are as follows:

(1) At the absorbance maxima of the class of compounds under investigation.
(2) Where appreciable absorbance occurs albeit not the maximum, such as 254 or 280 nm.
(3) In the region of high end absorption such as in the 190–230 nm range.

Table VII is a comprehensive list of spectral properties of some nucleosides, bases, and other biologically important compounds, normally found in serum or plasma ultrafiltrates or red blood cell lysates.

2. UV Spectra

In certain cases it may be advantageous to examine the entire spectrum of the eluting compound. Both on-line stopped-flow, fast-scanning UV detectors and photodiode array detectors have been used to obtain UV spectra (C1, D5, K32, S2).

In the stopped-flow UV scanning technique, it is usually possible to digitally store a background scan, thereby providing a means for obtaining background corrected spectra. Since solute diffusivity in the mobile phase is slow, solute diffusion effects can be neglected.

Figure 12 further illustrates the utility of the stopped-flow UV scanning technique. In this example, the absorption spectra of several peaks from the control serum sample (Fig. 10) are shown along with those of the reference spectra.

TABLE VII

Summary of Observed Retention Times, Peak-Height Ratios (280/254 nm), Fluorescence Response at 284 nm Excitation, 320 nm Cutoff

Compounds	Observed[a] retention time ± (T)	Peak-height ratio (1σ): 280 nm/254 nm	Fluorescence[b]
Cytosine (Cyt)	2.28 ± 0.04	0.437 (0.028)	—
Orotidine (Ord)	3.25 ± 0.03	0.516 (0.024)	—
Creatinine (Crt)	3.59 ± 0.06	0.001 (*****)[c]	—
Uracil (Ura)	4.13 ± 0.04	0.118 (0.018)	—
5-Aminoimidazole-4-carboxamide (AICA)	4.31 ± 0.13	0.621 (0.014)	—
L-Tyrosine (L-Tyr)	5.40 ± 0.07	2.47 (0.181)	WK
1-Methyladenine (1MeAde)	5.77 ± 0.10	0.194 (0.064)	—
Cytidine (Cyd)	5.85 ± 0.10	0.683 (0.024)	—
Hypoxanthine (Hyp)	7.31 ± 0.12	0.010 (*****)	—
Guanine (Gua)	7.56 ± 0.06	0.532 (0.014)	—
Uridine (Urd)	8.27 ± 0.19	0.215 (0.012)	—
Xanthine (Xan)	9.53 ± 0.05	0.459 (0.001)	—
5-Aminoimidazole-4-carboxamide riboside (AICAR)	8.77 ± 0.17	0.529 (0.037)	—
7-Methylinosine (7MeIno)	9.46 ± 0.19	0.286 (0.013)	ST
DL-Kynurenine (DL-Kyn)	9.82 ± 0.23	0.036 (0.001)	—
L-Phenylalanine (L-Phe)	10.40 ± 0.16	0.001 (*****)	WK
Allopurinol (Alo)	10.52 ± 0.21	0.023 (0.011)	—
7-Methylxanthosine (7MeXao)	10.69 ± 0.21	0.707 (0.008)	WK
3′,5′-cyclic CMP (c-CMP)	10.76 ± 0.25	0.630 (0.032)	—
5-Methylcytidine (5-MeCyd)	10.79 ± 0.19	1.130 (0.028)	—
Purine (Pur)	11.20 ± 0.31	0.148 (0.007)	—
7-Methylguanosine (7-MeGuo)	11.82 ± 0.16	0.572 (0.005)	MD
Pyrimidine (Pyr)	11.92 ± 0.25	0.319 (0.025)	—
Xanthosine (Xao)	12.40 ± 0.59	0.512 (0.022)	—
3′,5′-cyclic UMP (c-UMP)	12.86 ± 0.07	0.171 (0.004)	—
β-Nicotinamide adenine dinucleotide (β-NAD)	13.00 ± 0.16	0.120 (0.006)	—
Inosine (Ino)	13.52 ± 0.16	0.092 (0.007)	—
Adenine (Ade)	13.54 ± 0.15	0.080 (0.039)	—
Guanosine (Guo)	14.40 ± 0.18	0.373 (0.002)	—
7-Methylguanine (7MeGua)	14.65 ± 0.17	1.120 (0.046)	—
3′,5′-cyclic GMP (c-GMP)	15.12 ± 0.07	0.397 (0.041)	—
N^2-Methylguanine (2MeGua)	15.22 ± 0.22	0.535 (0.017)	—
3′,5′-cyclic IMP (c-IMP)	15.92 ± 0.06	0.113 (0.038)	—
Hippuric acid (HipA)	15.98 ± 0.05	0.001 (*****)	—
Deoxythymidine (dThd)	17.29 ± 0.11	0.565 (0.026)	—
L-Tryptophan (L-Trp)	17.47 ± 0.16	1.430 (0.048)	ST
1-Methylinosine (1MeIno)	18.22 ± 0.14	0.171 (0.088)	—
6-Methylpurine (6MePur)	18.98 ± 0.19	0.063 (0.002)	—
2-Methyladenine (2MeAde)	19.13 ± 0.04	0.129 (0.005)	—

TABLE VII (Continued)

Compounds	Observed[a] retention time ± (*T*)	Peak-height ratio (1σ): 280 nm/254 nm	Fluo-res-cence[b]
N'-Methylguanosine (1MeGuo)	19.22 ± 0.18	0.373 (0.014)	—
Purine riboside (PurRibo)	19.69 ± 0.16	0.099 (0.021)	—
Tubercydin (Tub)	19.73 ± 0.15	0.968 (0.009)	—
7-Methyladenine (7MeAde)	19.92 ± 0.17	0.126 (0.011)	—
N^2-Methylguanosine (2MeGuo)	20.23 ± 0.17	0.441 (0.045)	—
Kynurenic acid (KynA)	20.80 ± 0.16	0.099 (0.010)	WK
Adenosine (Ado)	21.77 ± 0.22	0.084 (0.005)	—
Theobromine (Thb)	22.51 ± 0.13	1.270 (0.029)	—
3',5'-cyclic AMP (c-AMP)	23.47 ± 0.31	0.125 (0.014)	—
N^2,N^2-Dimethylguanosine ($2Me_2Guo$)	24.42 ± 0.15	0.476 (0.007)	—
6-Methyladenine (6MeAde)	24.90 ± 0.15	0.465 (0.006)	—
6-Methyladenosine (6MeAdo)	30.86 ± 0.27	0.640 (0.047)	—
Caffeine (Caf)	35.08 ±0.27	1.260 (0.056)	—

[a] Error limits calculated as $\pm S_f - t_{x,f}/\sqrt{n}$ at $\alpha = 0.10$; *T* is time (min).

[b] Fluorescence measurements at $\lambda_{\text{excitation}} = 280$ nm, $\lambda_{\text{emission}} = 320$ nm; WK, weakly fluorescent; MD, moderately fluorescent; ST, strongly fluorescent.

[c] The asterisks indicate that the numbers are too small to be accurately determined.

3. *Spectrofluorometric Detection*

The use fluorescence detection in the chromatographic analysis of biological matrices has found widespread applicability because of its inherent sensitivity and specificity (B25, H5, H7–H9, K23, K29, K32, Y1, Y2). Since fluorescence is more sensitive than UV spectroscopy, it can be used advantageously to analyze for naturally fluorescent compounds in the presence of nonfluorescing, UV-absorbing compounds. In such cases, fluorometric detection is more sensitive while at the same time possible interferences are eliminated.

Figure 13 illustrates the use of spectrofluorometric detection in the reversed-phase chromatographic analysis of the control serum sample originally shown in Fig. 10. The advantages of fluorometric detection in eliminating background and increasing selectivity are obvious.

If the fluorescence detector is equipped with a motorized wavelength drive, it is possible to obtain stopped-flow excitation or emission spectra which can also aid in the characterization of biological compounds.

4. *Enzymatic Peak Shift*

The enzymatic peak-shift technique has been widely used for the identification of compounds or classes of compounds in biological matrices

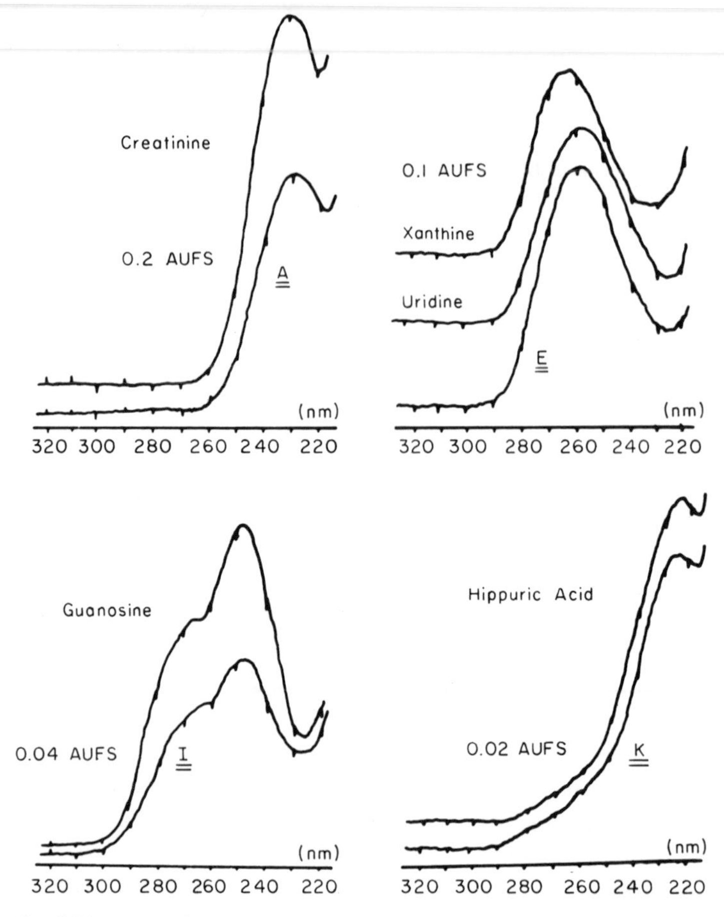

FIG. 12. UV spectra of several of the serum chromatographic peaks identified in Fig. 11, obtained using the stopped-flow technique. Sensitivities are indicated in the individual spectra. Reprinted with permission from Hartwick *et al.* (H7). Copyright by Elsevier Scientific Publishing Company.

(B27, B32, K28, K32, M22, R1). In addition to utilizing the sensitivity, specificity, and selectivity inherent in the enzymatic catalysis of reactions, the technique can be performed with a minimum amount of sample and sample workup, in a very short time. With the use of appropriate enzymes, it is possible to determine both the identity and purity of a substrate compound.

In the enzymatic peak-shift technique, the following criteria must be met:

(1) The enzyme for the catalytic reaction must be readily available.
(2) The substrate and/or products must show an appreciable response to the chromatographic detection system.
(3) The substrate and product must have different retention times.
(4) There must be no overlapping of retention times of the substrate and any other compound in the matrix.

Some of the common enzymes that have been used in the identification of nucleosides and bases are shown in Table VIII.

In this technique, an aliquot of the sample under investigation is first chromatographed. A second aliquot of sample is incubated with the enzyme under the appropriate conditions of pH and temperature. After a suitable time period, the incubated mixture is chromatographed. The disappearance of the substrate peak and/or appearance of product peaks confirms the identity and purity of the chromatographic peak. In cases where specific enzymes may not be available for a certain substrate, less specific enzymes, such as phosphatases, can be used for the identification of classes of compounds. In addition, it is possible to use coupled enzyme assays to drive a reaction to completion and thus characterize a peak in the chromatogram.

The use of the enzyme purine nucleoside phosphorylase for enzymatic peak identification is illustrated by the RPLC serum profile from a patient with severe depression. Based upon the retention time, the peak eluting at approximately 15 min was tentatively identified as inosine (Fig. 14A). Cochromatography with an inosine reference compound resulted in a subsequent increase in peak area for the compound of interest (Fig. 14B). Furthermore, stopped-flow UV spectra indicated a similarity between the inosine standard and the peak tentatively identified as inosine (Fig. 15).

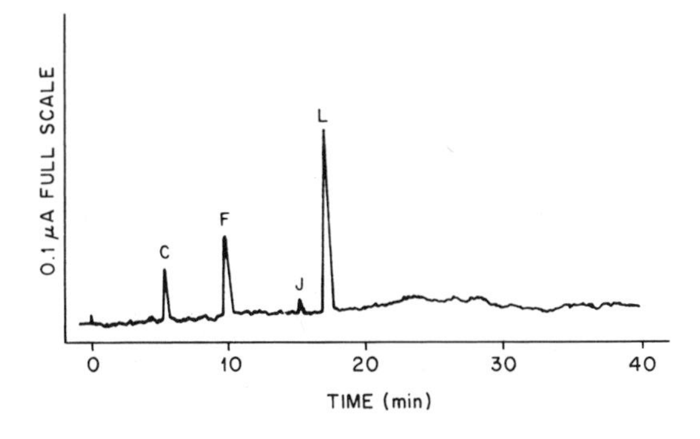

FIG. 13. Chromatogram of the control serum sample using spectrofluorometric detection (285 nm excitation, 320 nm cutoff). Conditions are the same as in Fig. 10. Reprinted with permission from Hartwick *et al.* (H7). Copyright by Elsevier Scientific Publishing Company, Amsterdam.

TABLE VIII

Enzymes Useful for the Identification of Some UV-Absorbing Compounds Found in Human Serum[a]

Substrate(s)	Reagent or co-factor	Enzyme	pH of reaction condition	Product(s)
Adenosine	H_2O	Adenosine deaminase (EC 3.5.4.4)	7.5	Inosine
Guanine	H_2O	Guanase (EC 3.4.4.3)	8.0	Xanthine
Guanosine	Phosphate	Phosphorylase (EC 2.4.2.1)	7.4	Guanine
Hypoxanthine, xanthine	H_2O, O_2	Xanthine oxidase (EC 1.2.3.2)	7.8	Xanthine, uric acid
Inosine	Phosphate	Purine nucleoside phosphorylase (EC 2.4.2.1)	7.4	Hypoxanthine, uric acid
L-Tryptophan	Pyridoxil 5-phosphate	Tryptophanase (EC 4.1.99.1)	8.3	Indole
Nucleoside 3′,5′-cyclic phosphate	Phosphate	Cyclic nucleotide phosphodiesterase (EC 3.1.4.17)	7.5	Nucleoside 5′-phosphate
Uric acid	H_2O, O_2	Uricase (EC 1.7.3.3)	8.5	Allantoin

[a] Reprinted with permission from Hartwick *et al.* (H7). Copyright by Elsevier Scientific Company, Amsterdam.

Finally, the serum sample was incubated with the enzyme purine nucleoside phosphorylase and rechromatographed (Fig. 14C). From the disappearance of the inosine peak and the appearance of a peak with the retention time of hypoxanthine, it can be concluded that the peak under investigation was indeed inosine.

5. *Chemical Derivatization and Characterization*

The use of organic and inorganic reagents has proved to be a useful peak identification technique in the chromatographic analysis of physiological fluids for nucleosides and bases (B25, H5, H7–H9, K32). Periodate oxidation has been used for the detection of cis-diol functionalities in ribonucleosides. The use of this reagent is illustrated in Fig. 16.

In this sample, the control serum sample in Fig. 10 has been reacted with sodium periodate and rechromatographed. The peaks H and I (in-

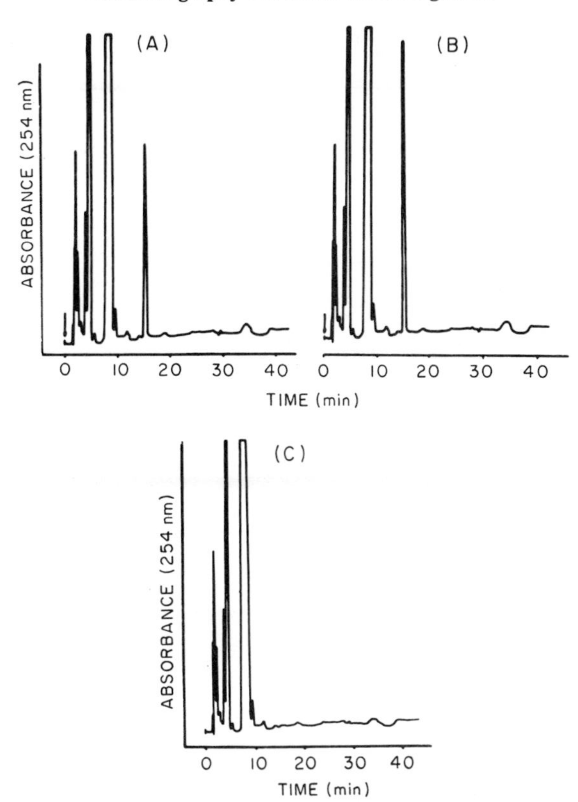

FIG. 14. (A) Chromatogram of a serum sample from a patient suffering from severe depression. (B) Chromatogram of the same serum sample co-injected with inosine. (C) Purine nucleoside phosphorylase peak shift of patient serum sample. Conditions are the same as in Fig. 10 except as follows: gradient: linear from 1 to 40% B in 35 min. Reprinted with permission from Krstulović *et al.* (K31). Copyright by Elsevier Scientific Publishing Company, Amsterdam.

osine and guanosine, respectively) have been removed, while the peak labeled E has been considerably reduced in area. It has been postulated, and since confirmed, that peak E is a combination of xanthine and uridine, of which the ribosyl moiety of uridine has been oxidized by the periodate.

A high degree of sensitivity and selectivity can be obtained with certain biomolecules by the chemical attachment of fluorophores. The most common fluorescent derivatization reagents include fluorescamine, dansyl chloride, pyridoxal, pyridoxal 5-phosphate, dansyl hydrazine, and pyridoxamine. Such derivatization procedures can be used to enhance the fluorescence of compounds with low quantum yields as well as impart fluorescent properties to compounds that do not fluoresce naturally.

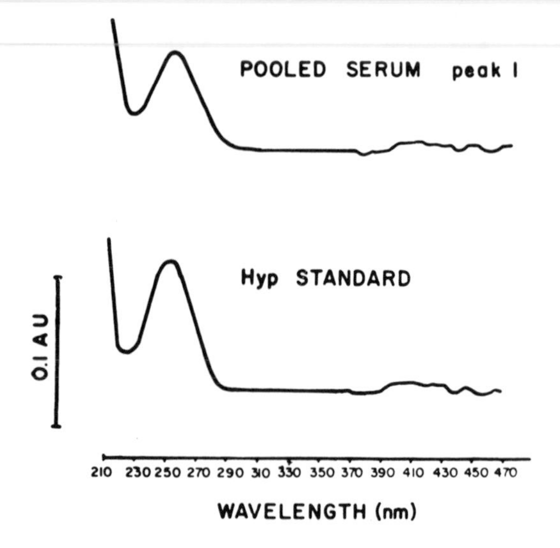

FIG. 15. Stopped-flow UV spectra of chromatographic peak in question from Fig. 14 and the inosine standard. Reprinted with permission from Krstulović *et al.* (K31). Copyright by Elsevier Scientific Publishing Company, Amsterdam.

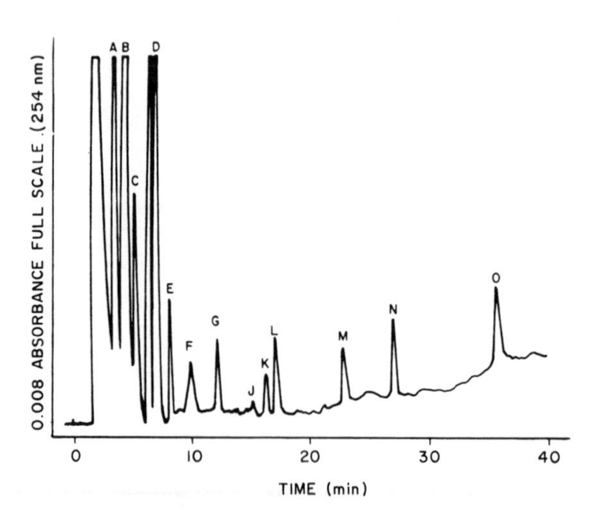

FIG. 16. Chromatogram of 80 μl of the control serum sample ultrafiltrate after reaction with 80 μl of 0.01 M $NaIO_4$. Total injection volume: 160 μl. Conditions are the same as in Fig. 10. Reprinted from permission from Hartwick *et al.* (H7). Copyright by Elsevier Scientific Publishing Company, Amsterdam.

V. BIOMEDICAL APPLICATIONS OF RPLC

A. Introduction

Reversed-phase high-performance liquid chromatography is a powerful analytical technique for the separation and quantitative analysis of many types of low-molecular-weight compounds in physiological fluids. The simultaneous identification and quantification of purine- and pyrimidine-based compounds in the cells and cellular fluids of individuals with disease or aberrations in metabolic pathways is useful in the basic understanding of disease mechanisms. Through the liquid chromatographic monitoring of major and modified nucleosides, nucleotides, and bases in physiological fluids, this technique has potential as a clinical diagnostic tool for the early detection of disease, to confirm a diagnosis, to monitor the course of a disease for early detection of recurrence, or to aid the clinician in planning effective therapy. In addition, RPLC can be used to monitor levels of nucleoside drugs and their metabolism in serum and plasma.

B. Metabolic Profiling

1. Nondiseased Population Groups

A comprehensive investigation of normal RPLC serum ribonucleoside and base profiles was conducted by Hartwick *et al.* (H7, H8), in which compounds normally found in serum were identified and quantified. Sample integrity was examined through the investigation of sample handling and storage procedures as well as prechromatographic processing techniques. The compounds identified include creatinine, uric acid, tyrosine, hypoxanthine, uridine, xanthine, inosine, guanosine, hippuric acid, and tryptophan, as well as the dietary compounds theobromine and caffeine. The RPLC serum profiles from normal subjects were consistent and reproducible and varied markedly from neoplastic serum profiles.

Karle *et al.* (K9) determined serum and plasma uridine levels in mice, rats, and humans by reversed-phase high-performance liquid chromatography. They report no difference in human plasma or serum uridine levels; however, there is some individual variation in human serum uridine levels throughout the day but no consistent pattern. From this study they concluded that uridine levels appear to be regulated and are not a direct reflection of dietary intake of uridine.

2. Diseased Population Groups

Realization of the biological significance to tRNAs has stimulated research directed at the elucidation of the many aspects of these biomolecules. The nucleic acid tRNA plays a key role in protein biosynthesis and regulatory activities in RNA metabolism and participates in the regulation of amino acid biosynthesis. The origin of tRNA in neoplastic cells was discussed by Borek and Kerr (B16), and it has been found that there is increased activity of tRNA methyltransferase enzymes in tumor cells (B10, F3, K10, M2, S21, S22, S33, S34). Although modified nucleosides have been found to occur in the urine of animals and humans from normal and diseased populations (M7), increased levels of these nucleosides indicate the extent of modification as well as the increased tRNA turnover rate (B15).

Increased levels of pseudouridine have been reported to be present in the urine of patients with various types of cancer (A1, M20, W1, W2). Since pseudouridine is only found in RNA, Kuo *et al.* (K38) developed a sensitive RPLC method for the rapid determination of urine pseudouridine levels. In a study of 10 colon cancer patients, they reported that 9 exhibited higher than normal pseudouridine-to-creatinine ratios. Davis *et al.* (D2) have also investigated urine ribonucleoside distribution patterns in patients with advanced colon cancer and report increased levels of 1-methylinosine, 1-methylguanosine, 2-methylguanosine, adenosine, and N^2,N^2-dimethylguanosine when compared to normal urine controls. Figure 17 illustrates the advanced colon cancer and normal urine chromatographic profiles.

Gehrke's group (G4) analyzed the urine of breast cancer and leukemia patients by RPLC. They report elevated levels of N^4-acetylcytidine in the breast cancer samples. They also report similar findings of elevated levels of N^2,N^2-dimethylguanosine, 1-methylguanosine, 1-methylinosine, and pseudouridine in the urine of patients with Burkitt's lymphoma, lung, colon, breast, and other types of cancer (W1, W2).

In the sera of patients with breast cancer, Krstulović *et al.* (K28) found 1-methylinosine and N^2,N^2-dimethylguanosine in increased concentrations. They also report on the presence of these compounds in the sera of patients with benign breast tumors, however, at much lower concentrations. Figure 18 illustrates nonneoplastic serum and breast cancer serum chromatographic profiles.

Moreover, in studies of patients with chronic lymphocytic leukemia and lung cancer, Hartwick *et al.* (H7) reported distinctive changes in their serum RPLC profiles vis-à-vis normal profiles.

Reversed-phase liquid chromatography has been used by Veening and co-workers (K22, S18) to monitor dialysate, serum, and urine from patients on artificial kidney machines. They report that RPLC is a reliable method for monitoring blood composition during dialysis.

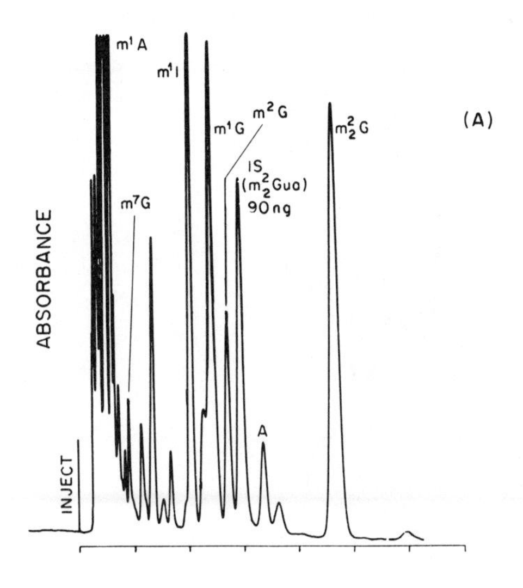

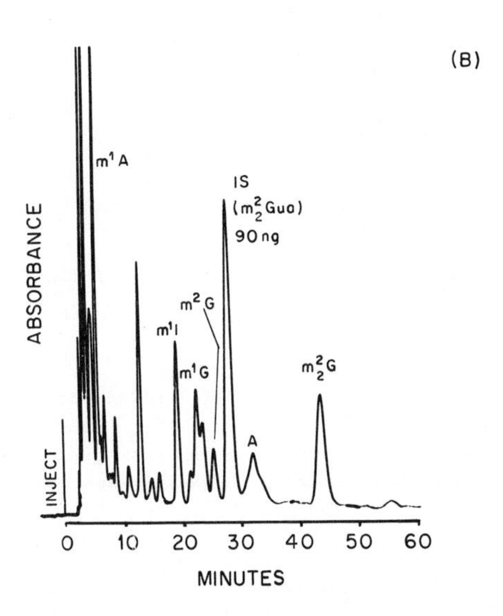

FIG. 17. (A) Reversed-phase HPLC isocratic separation of nucleosides in urine of colon cancer patient. (B) Reversed-phase HPLC isocratic separation of nucleosides in control urine. Sample: 50 μl equivalent to 25 μl urine; buffer: 0.05 M $NH_4H_2PO_4$, pH 5.10, with 50 ml methanol added per liter; flow rate: 1.0 ml/min; detector: 254 nm, 0.01 AUFS; temperature: 25°C. Reprinted with permission from Davis *et al.* (D2). Copyright by the American Association for Clinical Chemistry.

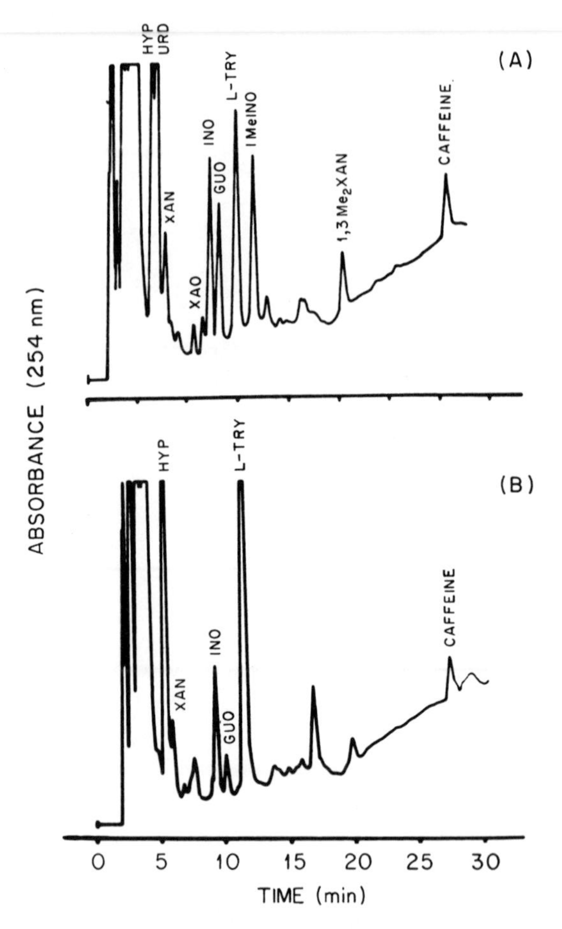

FIG. 18. (A) Chromatogram of a serum sample taken postoperatively from a nonfasting patient with breast cancer and metastatis to the bone. (B) Chromatogram of a serum sample from a normal, nonfasting subject. Chromatographic conditions same as in Fig. 14. Reprinted with permission from Krstulović *et al.* (K28). Copyright by Elsevier/North-Holland Press, Amsterdam.

C. Study of DNA and RNA Hydrolysates

Investigations of the nucleoside composition of nucleic acids have gained increased importance in recent years. A variety of high-performance liquid chromatographic methodologies have been developed for this purpose; however, none offer the advantages or versatility of the reversed-phase mode.

Mischke and Wickstrom (M17) developed an RPLC method for the analysis of deoxynucleosides in DNA and modified nucleosides in tRNAs. Using enzymatic hydrolysis techniques, they report that as little as 1–2 nmol of ribo- or deoxyribonucleosides can be determined.

Davis and co-workers (D1) also analyzed enzyme hydrolysates of tRNA by RPLC. They report excellent separation and resolution of both the major and modified ribonucleosides; however, enzyme hydrolysis may be less than desirable for quantitative release of ribonucleosides from tRNA. Figure 19 illustrates the tRNA hydrolysate separation of ribonucleosides in Hodgkin's tumor tRNA.

Major and modified deoxynucleosides were isolated from the enzymatic hydrolysis of DNA by Kuo *et al.* (K37). The chromatographic system uses a two-buffer step gradient with a reversed-phase C-18 column. This method has good sensitivity, selectivity, precision, and accuracy for the determination of all six deoxynucleosides without the use of drastic hydrolysis conditions or difficult sample preparation procedures. In addition, Nazar *et al.* (N3) and Salas and Sellinger (S1) have described liquid chromatographic methods for DNA and RNA hydrolysates.

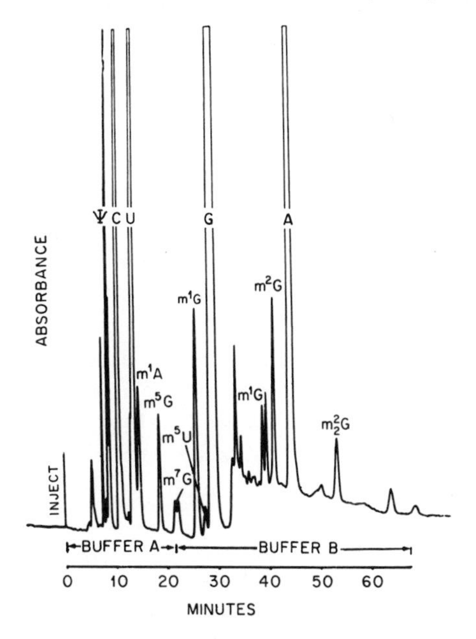

FIG. 19. Reversed-phase HPLC separation of nucleosides from tRNA hydrolysate. Sample: 20 μg Hodgkin's tumor tRNA; column: 600 × 4 mm μ-Bondapak C_{18}, buffers: (A) 2.5% (v/v) methanol in 0.01 *M* $NH_4H_2PO_4$, pH 5.10; (B) 10% (v/v) methanol in 0.01 *M* $NH_4H_2PO_4$, pH 5.10; flow rate; 1.0 ml/min; detector: 254 nm, 0.01 AUFS; temperature: 37.0°C. Reprinted with permission from Davis *et al.* (D1). Copyright by Elsevier Scientific Publishing Company, Amsterdam.

D. Enzyme Assays

Since many compounds can be analyzed simultaneously, a number of high-performance liquid chromatographic enzymatic assays have been developed in which both substrate(s) and product(s) can be monitored.

Pennington (P1) and Uberti (U1) first introduced the technique of liquid chromatographic enzyme assays by using the ion-exchange mode of HPLC in their analyses of 3′,5′-cyclic adenosine monophosphate phosphodiesterase and adenosine deaminase, respectively. Since that time, a number of liquid chromatographic enzyme assays have been developed.

Hartwick *et al.* (H11) used RPLC in developing an assay for adenosine deaminase which has been optimized for pH, ionic strength, reaction time, and substrate concentration. With this technique the substrate adenosine was monitored simultaneously with the products, hypoxanthine and inosine.

Krstulović and co-workers (K30) reported on the development of assays for serum acid and alkaline phosphatase enzymes. These enzymes have been reported to be elevated in various disease states (A2, B14, F2, K1). In addition, they developed a method for tryptophanase analysis using RPLC (K29). Employing fluorometric detection, the substrate, tryptophan, and reaction product, indole, can be monitored selectively with high sensitivity. They reported on several advantages of this method, including minimal sample preparation, rapid analysis time, and high specificity.

The most recent application of RPLC to the analysis of enzymes has been reported by Halfpenny and Brown (H1). An assay for purine nucleoside phosphorylase, a key mediator in the purine salvage pathway, has been developed and optimal conditions for the analysis determined. Figure 20 illustrates the simultaneous separation of the substrate, inosine, and products, uric acid and hypoxanthine. In another analysis, Halfpenny and Brown (H2) developed an assay for hypoxanthine-guanine phosphoribosyltransferase. Deficiency of this enzyme has been associated with Lesch–Nyhan syndrome as well as primary gout. The activity of the enzyme is determined by measurement of the decrease of the substrate, hypoxanthine, and increase in the product, inosine-5′-monophosphoric acid. A major advantage of using HPLC for enzyme assays is that the simultaneous measurement of both substrate and product reduces the error due to interference from competing enzymes.

E. Oligonucleotides

Recent interest in recombinant DNA techniques and the synthesis of biologically specific DNA have resulted in a number of high-performance

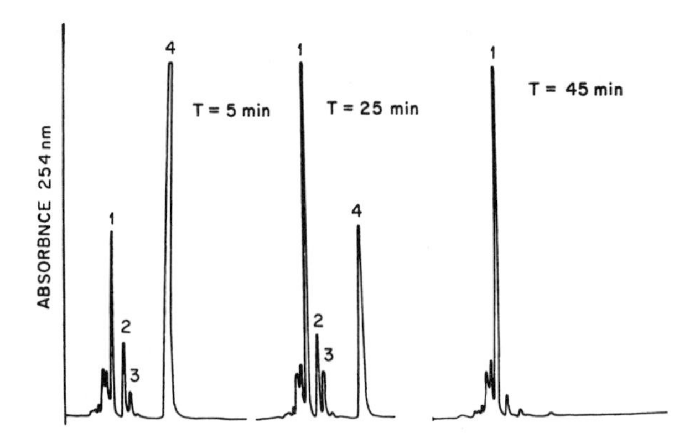

FIG. 20. Reaction of PNPase as a function of time. Chromatograms at various time intervals show the decrease of the substrate inosine (4) and the increase of the products uric acid (1), hypoxanthine (2), and xanthine (3). Reprinted with permission from Halfpenny and Brown (H2). Copyright by Elsevier Scientific Publishing Company, Amsterdam.

liquid chromatographic techniques for the isolation of synthetic oligonucleotide intermediates and the characterization of protected and unprotected oligo- and polynucleotides. Anion-exchange (B17, G1, G2, K11, S7, S8, V1), size exclusion (M18), polar bonded phases (C12), and mixed-mode ion exchange (E4, U3) have been successfully used for oligonucleotide analyses. In addition, reversed-phase HPLC has been used for the isolation and characterization of poly- and oligonucleotides (F5, G1, G2, J1, M6).

Fritz and co-workers (F5) described the use of reversed-phase HPLC for the isolation of protected oligonucleotide intermediates in the stepwise synthesis of deoxyribonucleotides. They report that the use of reversed-phase preparative scale HPLC reduces the time required for synthesis of oligonucleotides by 30%. In addition, they report on the RPLC separation of unprotected 5′-hydroxylated oligonucleotides.

Jones *et al.* (J1) have successfully used RPLC for the isolation and characterization of protected and unprotected mono- and oligonucleotides. In their synthetic sequence, the use of tert-butyldiphenylsilyl protecting groups allows for increased synthetic flexibility while at the same time increasing the hydrophobicity of oligonucleotides fragments, thereby allowing RPLC separation of oligonucleotide precursors and products.

The reversed-phase analysis of oligoribonucleotides has been reported by McFarland and Borer (M6). Employing linear gradient elution techniques, they have separated many dinucleotide monophosphates, homooligonucleotide mixtures, and block co-polymers. Crowther and

Hartwick (C12) also used RPLC to separate oligonucleotides and then used enzyme hydrolysis and subsequent RPLC chromatography of the nucleoside fragments for determination of the structures of the oligonucleotides.

VI. CONCLUSIONS

Reversed-phase high-performance liquid chromatography is an analytical technique of tremendous importance for the separation and quantitative analysis of compounds with widely varying chemical properties. Due to the versatility, sensitivity, and reproducibility of the technique, high-performance liquid chromatographic separations have become routine in biochemical and biomedical research, especially in investigations involving nucleic acid constituents.

To achieve increased resolution of closely related compounds, new column technologies have been introduced for the separation of nucleosides and their bases (A5, S11). The use of microbore columns will play an increasingly important role for achieving difficult separations or where sample size is limited (K34, R2, S12, S13, S15).

New detection systems will be developed for increased sensitivity which will place increased demands on prechromatographic sample treatment. Although many of the proposed liquid chromatographic/mass spectrometric interfaces have shown promise using model compounds (B3, B4, B11, B22, C2, C6, H17, K4, S3, S29, S30), their routine use for on-line characterization is not yet practical. Similar problems have been encountered with liquid chromatographic/infrared detection (K35, K36, T3). Detection systems such as electrochemical or radioactivity detectors will find increased use for the selective analysis or spectral characterization of eluting compounds (K24, M1, Y3, Y4).

The use of microprocessors for the control of the chromatograph components as well as for the acquisition and archival storage of data will be required for complex multicomponent analyses as well as to handle large numbers of samples and the large volumes of data generated.

Reversed-phase HPLC will find increased application in the analysis of purine antimetabolites and nucleoside antibiotics, both in the chemical laboratory for monitoring serum levels in chemotherapeutic treatment and in quality control in the pharmaceutical industry. In addition, RPLC will be used as a clinical diagnostic tool and aid the clinician in the detection of disease, in confirming a diagnosis, and in monitoring the causes of disease or effectiveness of therapy.

Acknowledgment

This work was supported, in part, by NCI Grant No. CA17603-06 (P.R.B.), American Hoechst Corporation, and Gillette Company Fellowships (H.A.S.).

REFERENCES

A1. W. S. Adams, F. Davis, and M. Nakatani, *Am. J. Med.* **28,** 726 (1960).
A2. A. C. Aisenberg, M. M. Kaplan, S. V. Rider, and J. M. Goldman, *Cancer* **26,** 318 (1970).
A3 F. S. Anderson and R. C. Murphy, *J. Chromatogr.* **121,** 251 (1976).
A4. N. G. Anderson, J. B. Green, M. L. Barber, and F. C. Ladd, *Anal. Biochem.* **6,** 153 (1963).
A5. S. P. Assenza and P. R. Brown, *J. Liq. Chromatogr.* **3,** 41 (1980).
B1. B. Bakay, E. Nissinen, and L. Sweetman, *Anal. Biochem.* **86,** 65 (1978).
B2. J. K. Baker, R. E. Skelton, and C.-Y. Ma, *J. Chromatogr.* **168,** 417 (1979).
B3. M. A. Baldwin and F. W. McLafferty, *Org. Mass Spectrom.* **7,** 1111 (1973).
B4. A. Bennington, A. Eicke, M. Junack, W. Sichtermann, J. Krizek, and H. Peters, *Org. Mass Spectrom.* **15,** 549 (1980).
B5. G. E. Berendsen and L. DeGalan, *J. Chromatogr.* **196,** 21 (1980).
B6. B. A. Bidlingmeyer, *J. Chromatogr. Sci.* **18,** 525 (1980).
B7. B. A. Bidlingmeyer, S. N. Deming, W. P. Price, Jr., B. Sachok, and M. Petrusek, *J. Chromatogr.* **186,** 419 (1979).
B8. B. A. Bidlingmeyer, S. N. Deming, W. P. Price, Jr., B. Sachok, and M. Petrusek, *in* "Advances in Chromatography, 1979" (A. Zlatkis, L. S. Ettre, and E. sz. Kováts, eds.), pp. 435–450. University of Houston, Houston, Texas, 1979.
B9. K. E. Bij, C. Horváth, W. R. Melander, and A. Nahum, *J. Chromatogr.* **203,** 65 (1981).
B10. B. B. Biswas, M. Edmonds, and R. Abrams, *Biochem. Biophys. Res. Comm.* **6,** 146 (1961).
B11. C. R. Blakely, J. J. Carmody, and M. L. Vestal, *Anal. Chem.* **52,** 1636 (1980).
B12. V. A. Bloomfield, D. M. Crothers, and I. Tinoco, Jr., "Physical Chemistry of Nucleic Acids," p. 53. Harper & Row, New York, 1974.
B13. V. A. Bloomfield, D. M. Crothers, and I. Tinoco, Jr., "Physical Chemistry of Nucleic Acids," p. 76. Harper & Row, New York, 1974.
B14. O. Bodansky, *in* "Enzymes in Health and Disease" (D. M. Greenberg and H. M. Harper, eds.), pp. 269–330. Thomas, Springfield, Illinois, 1960.
B15. E. Borek, B. S. Baliga, C. W. Gehrke, K. C. Kuo, S. Belman, W. Troll, and T. P. Waalkes, *Cancer Res.* **37,** 3362 (1977).
B16. E. Borek and J. J. Kerr, *Adv. Cancer Res.* **15,** 163 (1972).
B17. P. N. Borer, L. S. Kan, and P. O. P. Ts'o, *Biochemistry* **14,** 4847 (1975).
B18. K. O. Borg, M. Gabrielsson, and T. E. Jonsson, *Acta Pharm. Suec.* **11,** 313 (1974).
B19. H.-J. Breter, G. Seibert, and R. K. Zahn, *J. Chromatogr.* **140,** 251 (1977).
B20. H.-J. Breter and R. K. Zahn, *J. Chromatogr.* **137,** 61 (1977).
B21. A. D. Broom, M. P. Schweizer, and P. O. P. Ts'o, *J. Am. Chem. Soc.* **89,** 3612 (1967).
B22. J. J. Brophy, D. Nelson, and M. K. Withers, *Int. J. Mass Spectrom. Ion Phys.* **36,** 205 (1980).
B23. N. D. Brown, J. A. Kintzios, and S. E. Koetitz, *J. Chromatogr.* **177,** 170 (1979).

B24. P. R. Brown and A. M. Krstulović, "Reversed-phase High Performance Liquid Chromatography: Theory, Practice and Biomedical Applications." Wiley, New York, (1982).
B25. P. R. Brown, A. M. Krstulović, and R. A. Hartwick, *Adv. Chromatogr.* **18,** 101–138 (1980).
B26. P. R. Brown and E. Grushka, *Anal. Chem.* **52,** 1210 (1980).
B27. P. R. Brown, R. A. Hartwick, and A. M. Krstulović, *in* "Biological/Biomedical Applications of Liquid Chromatography" (G. N. Hawk, ed.), pp. 337–375. Marcel Dekker, Inc., New York, 1979.
B28. P. R. Brown, A. M. Krstulović, and R. A. Hartwick, *Hum. Hered.* **27,** 167 (1977).
B29. P. R. Brown, A. M. Krstulović, and R. A. Hartwick, *J. Clin. Chem. Clin. Biochem.* **14,** 282 (1976).
B30. P. R. Brown, S. Bobick, and F. L. Hanley, *J. Chromatogr.* **99,** 587 (1974).
B31. P. R. Brown, "High Pressure Liquid Chromatography: Biochemical and Biomedical Applications." Academic Press, New York, 1973.
B32. P. R. Brown, *J. Chromatogr.* **52,** 257 (1970).
B33. C. Bugge, *in* "The Jerusalem Symposium on Quantum Chemistry and Biochemistry," Vol. IV, pp. 194–195. Israel Academy of Sciences and Humanities, Jerusalem (1972).
B34. C. A. Burtis and D. Gere, "Nucleic Acid Constituents by Liquid Chromatography." Varian Aerograph, Walnut Creek, California, 1970.
B35. C. A. Burtis, M. N. Munk, and F. R. MacDonald, *Clin. Chem.* (*Winston-Salem, N.C.*) **16,** 201 (1970).
C1. C. D. Carr, *Anal. Chem.* **46,** 743 (1974).
C2. D. I. Carroll, I. Dzidic, R. N. Stillwell, K. D. Haegele, and E. C. Horning, *Anal. Chem.* **47,** 2369 (1975).
C3. S. Chen, D. M. Rosie, and P. R. Brown, *J. Chromatogr. Sci.* **15,** 218 (1977).
C4. F. K. Chow and E. Grushka, *J. Chromatogr.* **185,** 361 (1979).
C5. F. K. Chow and E. Grushka, *Anal. Chem.* **50,** 1346 (1978).
C6. R. G. Christensen, H. S. Hertz, S. Meiselman, and E. V. White, *Anal. Chem.* **53,** 171 (1981).
C7. M. B. Cohen, J. Maybaum, and W. Sadée, *J. Chromatogr.* **198,** 435 (1980).
C8. W. E. Cohn, *Science* **109,** 377 (1949).
C9. H. Colin and G. Guiochon, *J. Chromatogr.* **158,** 183 (1978).
C10. H. Colin, N. Ward, and G. Guiochon, *J. Chromatogr.* **149,** 169 (1978).
C11. N. H. C. Cooke, R. L. Viavattene, R. Eksteen, W. S. Wong, G. Davies, and B. L. Karger, *J. Chromatogr.* **149,** 391 (1978).
C12. J. B. Crowther and R. A. Hartwick, *J. Chromatogr.* **217,** 479 (1981).
D1. G. E. Davis, R. D. Suits, K. C. Kuo, and P. F. Agris, *J. Chromatogr.* **173,** 281 (1979).
D2. G. E. Davis, R. D. Suits, K. C. Kuo, C. W. Gehrke, T. P. Waalkes, and E. Borek, *Clin. Chem.* (*Winston-Salem, N.C.*) **23,** 1427 (1977).
D3. J. L. Day, L. Tterlikkis, R. Niemann, A. Mobley, and C. Spikes, *J. Pharm. Sci.* **67,** 1027 (1978).
D4. R. S. Deelder, H. A. J. Linssen, A. P. Konijnendijk, and J. L. M. Van de Venne, *J. Chromatogr.* **185,** 241 (1979).
D5. M. S. Denton, T. P. DeAngelis, A. M. Yacynych, W. R. Heineman, and T. W. Gilbert, *Anal. Chem.* **48,** 20 (1976).
D6. L. D'Souza and H. I. Glueck, *Thromb. Haemostatis* **38,** 990 (1977).
E1. M. Ehrlich and K. Ehrlich, *J. Chromatogr. Sci.* **17,** 531 (1979).
E2. S. Eksborg, P. O. Lagerström, R. Modin, and G. Schill, *J. Chromatogr.* **83,** 99 (1973).

E3. S. Eksborg and G. Schill, *Anal. Chem.* **45,** 2092 (1973).
E4. H. Eshaghpour and D. M. Crothers, *Nucleic Acids Res.* **5,** 13 (1979).
E5. J. M. Essigmann, W. F. Busby, Jr., and G. N. Wogan, *Anal. Biochem.* **81,** 384 (1977).
E6. J. E. Evans, H. Tiecklmann, E. W. Naylor, and R. Guthrie, *J. Chromatogr.* **163,** 29 (1979).
F1. A. F. Fell, S. M. Plag, and J. M. Neil, *J. Chromatogr.* **186,** 691 (1979).
F2. W. H. Fishman, N. I. Inglis, and M. J. Krant, *Clin. Chim. Acta* **12,** 298 (1965).
F3. E. Fleissner and E. Borek, *Biochemistry* **2,** 1093 (1963).
F4. A. Floridi, C. A. Palmerini, and C. Fini, *J. Chromatogr.* **138,** 203 (1977).
F5. H.-J. Fritz, R. Belagaje, E. L. Brown, R. H. Fritz, R. A. Jones, R. G. Less, and H. G. Khorana, *Biochemistry* **17,** 1257 (1978).
G1. M. J. Gait and R. C. Sheppard, *Nucleic Acids Res.* **4,** 4411 (1977).
G2. M. J. Gait and R. C. Sheppard, *Nucleic Acids Res.* **4,** 1135 (1977).
G3. C. W. Gehrke, K. C. Kuo, and R. W. Zumwalt, *J. Chromatogr.* **188,** 129 (1980).
G4. C. W. Gehrke, K. C. Kuo, G. E. Davis, R. D. Suits, T. P. Waalkes, and E. J. Borek, *J. Chromatogr.* **150,** 455 (1978).
G5. J. C. Giddings, "The Dynamics of Chromatography," Part I. Dekker, New York, 1965.
G6. C. Gilon, R. Leshem, Y. Tapuhi, and E. Grushka, *J. Am. Chem. Soc.* **101,** 7612 (1979).
G7. M. Glad, S. Ohlson, L. Hansson, M. Mansson, and K. Mosbach, *J. Chromatogr.* **200,** 254 (1980).
G8. E. Grushka and F. K. Chow, *J. Chromatogr.* **199,** 283 (1980).
H1. A. P. Halfpenny and P. R. Brown, *Pittsburgh Conf. Anal. Chem. App. Spectrosc., 1981* Abstract No. 285 (1981).
H2. A. P. Halfpenny and P. R. Brown, *J. Chromatogr.* **199,** 275 (1980).
H3. T. Hanai and K. Fujimura, *J. Chromatogr. Sci.* **14,** 140 (1976).
H4. P. E. Hare and E. Gil-Av, *Science* **204,** 1226 (1979).
H5. R. A. Hartwick and P. R. Brown, *CRC Crit. Rev. Anal. Chem.* **10,** 279 (1981).
H6. R. A. Hartwick, C. M. Grill, and P. R. Brown, *Anal. Chem.* **51,** 34 (1979).
H7. R. A. Hartwick, A. M. Krstulović, and P. R. Brown, *J. Chromatogr.* **186,** 659 (1979).
H8. R. A. Hartwick, S. P. Assenza, and P. R. Brown, *J. Chromatogr.* **186,** 647 (1979).
H9. R. A. Hartwick, S. P. Assenza, and P. R. Brown, *in* "Advances in Chromatography, 1979" (A. Zlatkis, L. S. Ettre, and E. sz. Kováts, eds.), pp. 725–736. University of Houston, Houston, Texas, 1979.
H10. R. A. Hartwick, D. A. Van Haverbeke, M. McKeag, and P. R. Brown, *J. Liq. Chromatogr.* **2,** 725 (1979).
H11. R. A. Hartwick, A. Jefferies, A. M. Krstulović, and P. R. Brown *J. Chromatogr. Sci.* **16,** 427 (1978).
H12. R. A. Hartwick and P. R. Brown, *J. Chromatogr.* **143,** 383 (1977).
H13. R. A. Hartwick and P. R. Brown, *J. Chromatogr.* **126,** 679 (1976).
H14. R. A. Hartwick and P. R. Brown, *J. Chromatogr.* **112,** 651 (1975).
H15. H. Hemetsberger, M. Kellerman, and H. Ricken, *Chromatographia* **10,** 276 (1977).
H16. H. Hemetsberger, W. Maasfeld, and H. Ricken, *Chromatographia* **9,** 303 (1976).
H17. J. D. Henion, *Anal. Chem.* **50,** 1687 (1978).
H18. R. A. Henry, J. A. Schmit, and R. C. Williams, *J. Chromatogr. Sci.* **11,** 358 (1972).
H19. N. E. Hoffman and J. C. Liao, *Anal. Chem.* **49,** 2231 (1977).
H20. M. Hori, *in* "Methods in Enzymology" Vol. 7, p. 381. Academic Press, New York, 1967.
H21. C. Horváth, W. Melander, and A. Nahum, *J. Chromatogr.* **186,** 371 (1979).

H22. C. Horváth and W. R. Melander, *Am. Lab.* **10,** 17 (1978).
H23. C. Horváth and W. R. Melander, *J. Chromatogr. Sci.* **15,** 393 (1977).
H24. C Horváth, W. Melander, I. Molnár, and P. Molnár, *Anal. Chem.* **49,** 2295 (1977).
H25. C. Horváth, W. Melander, and I. Molnár, *Anal. Chem.* **49,** 142 (1977).
H26. C. Horváth, W. Melander, and I. Molnár, *J. Chromatogr.* **125,** 129 (1976).
H27. C. Horváth and S. R. Lipsky, *Anal. Chem.* **41,** 1227 (1969).
H28. D.-S. Hsu and S. S. Chen, *J. Chromatogr.* **192,** 193 (1980).
H29. P. Hubert and J. Porath, *J. Chromatogr.* **206,** 164 (1981).
J1. R. A. Jones H.-J. Fritz, and H. G. Khorana, *Biochemistry* **17,** 1268 (1978).
J2. E. Juengling and K. Kammermeier, *Anal. Biochem.* **102,** 358 (1980).
K1. M. M. Kaplan and L. Rogers, *Lancet* **2,** 1029 (1969).
K2. K. Karch, I. Sabestian, I. Halász, and H. Engelhardt, *J. Chromatogr.* **122,** 171 (1976).
K3. B. L. Karger, J. N. LePage, and N. Tanaka, *in* "High Performance Liquid Chromatography: Advances and Perspectives" (C. Horváth, ed.), Vol. 1, pp. 113–206. Academic Press, New York, 1980.
K4. B. L. Karger, D. P. Kirby, P. Vouros, R. L. Foltz, and B. Hidy, *Anal. Chem.* **51,** 2324 (1979).
K5. B. L. Karger and R. W. Giese, *Anal. Chem.* **50,** 1048A (1978).
K6. B. L. Karger, J. R. Grant, A. Hartkopf, and P. H. Weiner, *J. Chromatogr.* **128,** 65 (1976).
K7. B. L. Karger and B. A. Persson, *J. Chromatogr. Sci.* **12,** 521 (1974).
K8. B. L. Karger, L. R. Snyder, and C. Horváth, "An Introduction to Separation Science." Wiley, New York, 1973.
K9. J. M. Karle, L. W. Anderson, D. D. Dietrick, and R. L. Cysyk, *Anal. Biochem.* **109,** 41 (1980).
K10. S. J. Kerr, *Cancer Res.* **35,** 2969 (1975).
K11. H. G. Khorana and J. P. Vizsolyi, *J. Am. Chem. Soc.* **83,** 675 (1961).
K12. J. X. Khym, *Clin. Chem. (Winston-Salem, N.C.)* **21,** 1245 (1975).
K13. J. J. Kirkland, *J. Chromatogr. Sci.* **10,** 129 (1972).
K14. J. J. Kirkland, *J. Chromatogr. Sci.* **9,** 206 (1971).
K15. P. T. Kissinger, *Anal. Chem.* **49,** 883 (1977).
K16. J. H. Knox and R. A. Hartwick, *J. Chromatogr.* **204,** 3 (1981).
K17. J. H. Knox and J. Jurand, *J. Chromatogr.* **203,** 85 (1981).
K18. J. H. Knox, *J. Chromatogr. Sci.* **15,** 352 (1977).
K19. J. H. Knox and J. Jurand, *J. Chromatogr.* **125,** 89 (1976).
K20. J. H. Knox and G. R. Laird, *J. Chromatogr.* **122,** 17 (1976).
K21. J. H. Knox and A. Pyrde, *J. Chromatogr.* **112,** 171 (1975).
K22. E. J. Knudson, Y. C. Lauo, H. Veening, and D. A. Dayton, *Clin. Chem. (Winston-Salem, N.C.)* **24,** 686 (1978).
K23. S. Kobayashi and K. Imai, *Anal. Chem.* **52,** 424 (1980).
K24. R. C. Kong, B. Sachok, and S. N. Deming, *J. Chromatogr.* **199,** 307 (1980).
K25. A. P. Konijnendijk and J. L. M. Van de Venne, *in* "Advances in Chromatography, 1979" (A. Zlatkis, L. S. Ettre, and E. sz. Kováts, eds.), pp. 451–462. University of Houston, Houston, Texas, 1979.
K26. J. C. Kraak, K. M. Jonker, and J. F. K. Huber, *J. Chromatogr.* **142,** 671 (1977).
K27. G. Krauss and H. Reinboth, *Anal. Biochem.* **78,** 1 (1977).
K28. A. M. Krstulović, R. A. Hartwick, and P. R. Brown, *Clin. Chim. Acta* **97,** 159 (1979).
K29. A. M. Krstulović and C. Matzura, *J. Chromatogr.* **176,** 217 (1979).
K30. A. M. Krstulović, R. A. Hartwick, and P. R. Brown, *J. Chromatogr.* **163,** 19 (1978).

K31. A. M. Krstulović, R. A. Hartwick, P. R. Brown, and K. Lohse, *J. Chromatogr.* **158,** 365 (1978).
K32. A. M. Krstulović, P. R. Brown, and D. M. Rosie, *Anal. Chem.* **49,** 2237 (1977).
K33. A. M. Krstulović, P. R. Brown, D. M. Rosie, and P. B. Champlin, *Clin. Chem. (Winston-Salem, N.C.)* **23,** 1984 (1977).
K34. P. Kucera, *J. Chromatogr.* **198,** 93 (1980).
K35. D. Kuehl and P. R. Griffiths, *Anal. Chem.* **52,** 1394 (1980).
K36. D. Kuehl and P. R. Griffiths, *J. Chromatogr. Sci.* **17,** 471 (1979).
K37. K. C. Kuo, R. A. McCune, C. W. Gehrke, R. Midgett, and M. Ehrlich, *Nucleic Acids Res.* **8,** 4763 (1980).
K38. K. C. Kuo, C. W. Gehrke, R. A. McCune, T. P. Waalkes, and E. Borek, *J. Chromatogr.* **145,** 383 (1978).
L1. D. P. Lee and J. H. Kindsvater, *Anal. Chem.* **52,** 2425 (1980).
L2 C. P. H. Leigh and P. J. Cashion, *J. Chromatogr.* **192,** 490 (1980).
L3. R. E. Leitch and J. J. DeStafano, *J. Chromatogr. Sci.* **11,** 105 (1973).
L4. A. Leyva, J. Schornagel, and H. M. Pinedo, *J. Clin. Chem. Clin. Biochem.* **17,** 422 (1979).
L5. W. Lindner, J. N. LePage, G. Davies, D. E. Seitz, and B. L. Karger, *J. Chromatogr.* **185,** 323 (1979).
L6. C. H. Lochmüller and D. R. Wilder, *J. Chromatogr. Sci.* **17,** 574 (1979).
L7. D. C. Locke, *J. Chromatogr. Sci.* **12,** 433 (1974).
L8. C. D. Lothrop, Jr. and M. Uziel, *Anal. Biochem.* **109,** 160 (1980).
L9. C. Lush, Z. H. A. Rahim, D. Perrett, and J. R. Griffiths, *Anal. Biochem.* **93,** 227 (1979).
M1. D. J. Malcolme-Lawes and P. Warwick, *J. Chromatgr.* **200,** 47 (1980).
M2. L. R. Mandel and E. Borek, *Biochemistry* **2,** 555 (1963).
M3. J. Maybaum, F. K. Klein, and W. Sadee, *J. Chromatogr.* **188,** 149 (1980).
M4. A. McBurney and T. Gibson, *Clin. Chim. Acta* **102,** 19 (1980).
M5. R. M. McCormick and B. L. Karger, *J. Chromatogr.* **199,** 259 (1980).
M6. G. D. McFarland and P. N. Borer, *Nucleic Acids Res.* **7,** 1067 (1979).
M7. E. S. McFarlane and G. J. Shaw, *Can. J. Microbiol.* **14,** 185 (1968).
M8. M. McKeag and P. R. Brown, *J. Chromatogr.* **152,** 253 (1978).
M9. W. R. Melander and C. Horváth *in* "High Performance Liquid Chromatography: Advances and Perspectives" (C. Horváth, ed.), Vol. 2, pp. 113–319. Academic Press, New York, 1980.
M10. W. R. Melander, K. Kalghatgi, and C. Horváth, *J. Chromatogr.* **201,** 201 (1980).
M11. W. R. Melander and C Horváth, *J. Chromatogr.* **201,** 211 (1980).
M12. W. R. Melander, J. Stoveken, and C. Horváth, *J. Chromatogr.* **199,** 35 (1980).
M13. W. R. Melander, A. Nahum, and C. Horváth, *J. Chromatogr.* **185,** 129 (1979).
M14. W. R. Melander, J. Stoveken, and C. Horváth, *J. Chromatogr.* **185,** 111 (1979).
M15. W. R. Melander, D. E. Campbell, and C. Horváth, *J. Chromatogr.* **158,** 215 (1978).
M16. J. M. Miller, "Separation Methods in Chemical Analysis." Wiley, New York, 1975.
M17. C. F. Mischke and E. Wickstrom, *Anal. Biochem.* **105,** 181 (1980).
M18. D. Molko, R. Derbyshire, A. Guy, A. Roget, R. Teoule, and A. Boucherle, *J. Chromatogr.* **206,** 493 (1981).
M19. R. G. Moran and W. C. Werkheiser, *Anal Biochem.* **88,** 668 (1978).
M20. J. Mrochek, S. Dinsmore, and T. P. Waalkes, *JNCI, J. Natl. Cancer Inst.* **53,** 1553 (1974).
M21. F. Murakami, S. Rokushika, and H. Hatano, *J. Chromatogr.* **53,** 584 (1970).
M22. G. C. Mylls, F. C. Schmalstieg, K. B. Trimmer, A. S. Goldman, and R. M. Goldblum, *Proc. Natl. Acad. Sci. U.S.A.* **73,** 2867 (1976).

N1. A. Nahum and C. Horváth, *J. Chromatogr.* **203,** 53 (1981).
N2. R. N. Nazar, W. H. Spohn, and H. Busch, *Anal. Biochem.* **74,** 615 (1976).
N3. E. Nissinen, *Anal. Biochem.* **106,** 497 (1980).
O1. J. B. Opienska, M. Charezinski, and H. Bruszkiewicz, *Clin. Chim. Acta* **8,** 206 (1963).
P1. S. M. Pennington, *Anal. Chem.* **43,** 1701 (1971).
P2. E. H. Pfadenhauer and S. D. Tong, *J. Chromatogr.* **162,** 585 (1979).
P3. W. Plunkett, J. A. Benvenuto, D. J. Stewart, and T. L. Lod, *Cancer Treat. Rep.* **63,** 415 (1979).
P4. R. T. Pon and K. K. Ogilvie, *J. Chramatogr.* **205,** 202 (1981).
R1. H. Ratech, G. J. Thorbecke, and R. Hirschhorn, *J. Chromatogr.* **183,** 499 (1980).
R2. C. E. Reese and R. P. W. Scott, *J. Chromatogr. Sci.* **18,** 479 (1980).
R3. T. L. Riss, N. L. Zorich, M. D. Williams, and A. Richardson, *J. Liq. Chromatogr.* **3,** 133 (1980).
R4. E. J. Ritter and L. M. Bruce, *Biochem. Med.* **21,** 16 (1979).
R5. Y. M. Rustum, *Anal. Biochem.* **90,** 3289 (1978).
S1. C. E. Salas and O. Z. Sellinger, *J. Chromatogr.* **133,** 231 (1977).
S2. R. E. Santini, M. J. Milano, and H. L. Pardue, *Anal. Chem.* **45,** 915A (1973).
S3. K. H. Schafer and K. Levsen, *J. Chromatogr.* **206,** 245 (1981).
S4. G. Schill, *Acta Pharm. Suec.* **2,** 13 (1965).
S5. G. Schill, R. Modin, and B. Persson, *Acta Pharm. Suec.* **2,** 119 (1965).
S6. G. Schöch, J. Thomale, H. Lorenz, H. Suberg, and U. Karsten, *Clin. Chim. Acta* **108,** 247 (1980).
S7. H. Schott, *J. Chromatogr.* **172,** 179 (1979).
S8. H. Schott and M. Schwarz, *J. Chromatogr.* **157,** 197 (1978).
S9. P. D. Schweinsberg and T. L. Lou, *J. Chrmatogr.* **181** 103 (1980).
S10. M. P. Schweizer, S. I. Chan, and P. O. P. Ts'o, *J. Am. Chem. Soc.* **87,** 5241 (1965).
S11. H. A. Scoble, S. P. Assenza, and P. R. Brown, *Pittsburgh Conf. Anal. Chem. Appl. Spectrosc., 1980* Abstract No. 674.
S12. R. P. W. Scott, P. Kucera, and M. Munroe, *J. Chromatogr.* **186,** 475 (1979).
S13. R. P. W. Scott and P. Kucera, *J. Chromatogr.* **185,** 27 (1979).
S14. R. P. W. Scott and P. Kucera, *J. Chromatogr.* **175,** 51 (1979).
S15. R. P. W. Scott and P. Kucera, *J. Chromatogr.* **169,** 51 (1979).
S16. R. P. W. Scott and P. Kucera, *J. Chromatogr.* **142,** 213 (1979).
S17. R. P. W. Scott, "Contemporary Liquid Chromatography." Wiley, New York, 1976.
S18. F. C. Senftleber, A. G. Halline, H. Veening, and D. A. Dayton, *Clin. Chem. (Winston-Salem, N.C.)* **22,** 1522 (1976).
S19. R. Seuffer, W. Voelter, and W. Bauer, *J. Clin. Chem. Clin. Biochem.* **15,** 663 (1977).
S20. B. Shaikh, S.-S. Huang, N. J. Pontzer, and W. L. Zielinski, Jr., *J. Liq. Chromatogr.* **1,** 75 (1978).
S21. O. K. Sharma, S. J. Kerr, P. Lipshitz-Wiesner, and E. Borek, *Fed. Proc., Fed. Am. Soc. Exp. Biol.* **30,** 167 (1971).
S22. B. Sheid, T. Lu, and J. H. Nelson, *Cancer Res.* **34,** 2416 (1974).
S23. C. S. Simpson, "Practical High Performance Liquid Chromatography," Part I. Heyden, London, 1976.
S24. R. P. Singhal, R. K. Bajaj, C. M. Buess, D. B. Smoll, and V. N. Vakharia, *Anal. Biochem.* **109,** 1 (1980).
S25. R. P. Singhal, *Sep. Purif. Methods* **3,** 339 (1974).
S26. R. P. Singhal and W. E. Cohn, *Biochemistry* **12,** 1532 (1973).
S27. R. P. Singhal, *Arch. Biochem. Biophys.* **152,** 800 (1972).

S28. M. A. Slifkin, *in* "Physicochemical Properties of Nucleic Acids" (J. Duschesne, ed.), Vol. 1, p. 70. Academic Press, New York, 1973.
S29. R. D. Smith and A. L. Johnson, *Anal. Chem.* **53,** 1120 (1981).
S30. R. D. Smith and A. L. Johnson, *Anal. Chem.* **53,** 739 (1981).
S31. T. N. Solie and J. A. Schellman, *J. Mol. Biol.* **33,** 61 (1968).
S32. S. P. Sood, L. E. Sartori, D. P. Wittmer, and W. G. Haney, Jr., *Anal. Chem.* **48,** 796 (1976).
S33. P. R. Srinivasan and E. Borek, *Proc. Natl. Acad. Sci. U.S.A.* **49,** 529 (1963).
S34. J. L. Starr, *Biochim. Biophys. Acta* **61,** 676 (1962).
T1. G. A. Taylor, P. J. Dady, and K. R. Harrap, *J. Chromatogr.* **183,** 421 (1980).
T2. M. J. Telepchak, *Chromatographia* **6,** 234 (1973).
T3. N. Teramae and S. Tanaka, *Spectrosc. Lett.* **13,** 117 (1980).
T4. C. P. Terweij-Groen, S. Heemstra, and J. C. Kraak, *J. Chromatogr.* **161,** 69 (1978).
T5. P. O. P. Ts'o, N. S. Kondo, R. K. Robbins, and A. D. Broom, *J. Am. Chem. Soc.* **91,** 5625 (1969).
T6. P. O. P. Ts'o, *in* "Molecular Associations in Biology" (B. Pullman, ed.), p. 59. Academic Press, New York, 1968.
T7. P. O. P. Ts'o and S. I. Chan, *J. Am. Chem. Soc.* **86,** 4176 (1964).
T8. P. O. P. Ts'o, I. S. Melvin, and A. Olsen, *J. Am. Chem. Soc.* **85,** 1289 (1963).
U1. J. Uberti, J. J. Lightbody, and R. M. Johnson, *Anal. Biochem.* **80,** 1 (1977).
U2. K. K. Unger, N. Becker, and P. Roumeliotis, *J. Chromatogr.* **125,** 115 (1976).
U3. D. A. Usher, *Nucleic Acids Res.* **6,** 2289 (1979).
U4. M. Uziel, L. H. Smith, and S. A. Taylor, *Clin. Chem.* (*Winston-Salem, N.C.*) **22,** 1451 (1976).
U5. M. Uziel, C. K. Koh, and W. E. Cohn, *Anal. Biochem.* **25,** 77 (1968).
V1. A. Vandenberghe, L. Nelles, and R. DeWachter, *Anal. Biochem.* **107,** 369 (1980).
V2. J. L. M. Van de Venne, J. L. H. M. Hendrikx, and R. S. Deelder, *J. Chromatogr.* **167,** 1 (1978).
V3. D. A. Van Haverbeke and P. R. Brown, *J. Liq. Chromatogr.* **1,** 507 (1978).
V4. W. Voelter, K. Zech, P. Arnold, and G. Ludwig, *J. Chromatogr.* **199,** 345 (1980).
V5. W. Voelter and H. Bauer, *Clin. Chem.* (*Winston-Salem, N.C.*) **21,** 1882 (1975).
W1. T. P. Waalkes, C. W. Gehrke, W. A. Bleyer, R. W. Zumwalt, C. L. M. Olweny, K. C. Kuo, D. B. Lakings, and S. A. Jacobs, *Cancer Chemother. Rep.* **59,** 721 (1975).
W2. T. P. Waalkes, C. W. Gehrke, R. W. Zumwalt, S. Y. Chang, D. B. Lakings, D. L. Ahmann, and C. G. Moertel, *Cancer* **36,** 390 (1975).
W3. T. F. Walseth, G. Graaf, M. C. Moos, Jr., and N. D. Goldberg, *Anal. Biochem.* **107,** 240 (1980).
W4. D. P. Wittmer, N. O. Nuessle, and W. G. Haney, Jr., *Anal. Chem.* **47,** 1422 (1975).
W5. W. E. Wung and S. B. Howell, *Clin. Chem.* (*Winston-Salem, N.C.*) **26,** 1704 (1980).
Y1. E. S. Yeung and M. J. Sepaniak, *Anal. Chem.* **52,** 1465A (1980).
Y2. E. S. Yeung, L. E. Steenhoek, S. D. Woodruff, and J. C. Kuo, *Anal. Chem.* **52,** 1399 (1980).
Y3. R. Yost, J. Stoveken, A. F. Poile and W. MacLean, *Chromatogr. Newslet.* (Perkin-Elmer) **5,** 28 (1977).
Y4. R. Yost, J. Stoveken, and W. MacLean, *J. Chromatogr.* **134,** 73 (1977).
Z1. M. Zakaria, P. R. Brown, and E. Grushka, *in* "Advances in Chromatography, 1981" (A. Zlatkis, R. Segura, and L. S. Ettre, eds.), pp. 451–471. University of Houston, Houston, Texas, 1981.

THE SEPARATION OF PROTEINS BY REVERSED-PHASE HIGH-PERFORMANCE LIQUID CHROMATOGRAPHY

William S. Hancock

Department of Chemistry, Biochemistry, and Biophysics
Massey University
Palmerston North, New Zealand

and

James T. Sparrow

Department of Medicine
Baylor College of Medicine
and The Methodist Hospital
Texas Medical Center
Houston, Texas

HIGH-PERFORMANCE LIQUID
CHROMATOGRAPHY, Vol. 3

ISBN 0-12-312203-1

I. INTRODUCTION

A. Specific Problems Associated with Separation of Proteins by Reversed-Phase High-Performance Liquid Chromatography

Hydrophobic chromatography has become a popular technique for the purification of proteins on alkyl- or aryl-substituted agarose (*27, 45*). The use of the technique has been limited by low efficiencies obtained with these agarose derivatives, slow separations, and the possibility of protein denaturation by solvents or longer alkyl-chain substituents (*26, 31*). These disadvantages have not prevented a number of excellent separations by hydrophobic chromatography, and thus the purification of proteins by reversed-phase high-performance liquid chromatography (HPLC) is technically feasible.

Although the separation of amino acids and small peptides by reversed-phase HPLC is becoming an accepted procedure, the application of this technique to the purification of proteins still requires careful evaluation. The biological activities of many proteins are sensitive to denaturation by extremes in pH, by contact with organic solvents or high salt concentrations, by adsorption into glass or hydrophobic moieties, or at an air–water interface (*21*).

An early concern with the HPLC technique was the use of high pressures to achieve high flow rates of the mobile phase through a column packed with microparticulate silica. Recent improvements in column design and operating procedures, however, allow the purification of proteins at modest pressures (e.g., 500 psi) and flow rates (30–60 ml/h). Since it has been reported that C_3-alkyl chains are compatible with catalytic activity of adsorbed and eluted proteins, but larger alkyl substituents may cause denaturation (*26*), the use of reversed-phase columns of medium polarity, e.g., —C_3H_7-phenyl, when combined with a judicious choice of organic modifier and salt concentrations (e.g., isopropanol and phosphate) at pH 3–7, should allow the separation of many proteins with retention of biological activity.

B. The Separation Mechanism

Most organic molecules which occur in biological systems can be characterized as polar molecules with a large number of charged groups, i.e., $—NH_3^+$, $—COO^-$, and —OH groups. For this reason, reversed-phase chromatography with an aqueous mobile phase is particularly suitable for the analysis of such molecules. The selection of the optimal separation system can be a complex process because of the variety of interactions which may occur between the sample molecule, solvent, and stationary phase. For this reason it is necessary to consider the possible mechanisms by which a reversed-phase separation may occur. Liquid chromatographic theory has been developed largely from experience gained in the analysis of low-molecular-weight molecules, first by gas–liquid and then by liquid–liquid chromatography. Such experience, however, may not be directly applicable to the separation of biological macromolecules, which have higher molecular weights, many charged functional groups, and conformational flexibility.

Two important hypotheses which describe interactions between proteins and hydrophobic systems are the fluid mosaic model for membrane structure proposed by Singer and Nicholson (*60*) and the amphipathic apolipoprotein structure proposed by Segrest *et al.* (*56*). In Singer's model for membrane structure shown in Fig. 1, it is proposed that membrane proteins float in a sea of lipid. The nonpolar regions of the protein are in contact with the nonpolar lipids present in the membrane, whereas the polar surface is exposed to the aqueous environment.

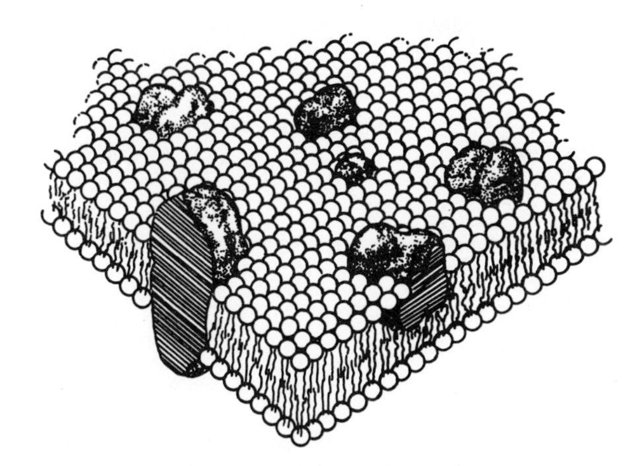

FIG. 1. The lipid–globular protein fluid mosaic model of membrane structure. The solid bodies represent globular proteins, some of which span the phospholipid bilayer represented by the open circles. Adapted from Singer and Nicholson (1972).

A variety of studies have suggested that in the presence of lipid, the apoprotein components of lipoproteins assume regions of high helical content which exhibit two faces of opposite polarity. When these proteins are on the surface of a lipoprotein particle, the nonpolar face can interact with the nonpolar lipid components of the particle, whereas the polar face interacts with the polar head groups of cholesterol and phospholipid, and the aqueous environment of the bloodstream (Fig. 2).

As is common for many biological systems, the hydrophobic effect plays an important role in both models. The effect contributes a favorable energetic state to the close association of nonpolar (hydrophobic) surfaces in aqueous environments. In the absence of such an association, the nonpolar surfaces must be exposed to an aqueous environment and considerable energy is required to form a solvent cavity for the molecules. There may well be a useful parallel between these models and the mechanism of separation of proteins by reversed-phase HPLC.

In Fig. 3, a model is shown for the interaction between protein molecules and a nonpolar support. Unlike the lipid present in membranes and lipoprotein particles, the reversed-phase C_{18} group is anchored to the polar silica surface. It is not possible, therefore, for the C_{18} chains to completely avoid the polar silica surface and the polar mobile phase. In 1972, Scott and Kucera (*55*) provided evidence that the nonpolar regions of the organic modifier are oriented in a manner that shields the C_{18} phase from both polar phases. In Fig. 3, the modifier is shown as methanol, but a similar result would be expected for other solvents such as CH_3CN, $(CH_3)_2CHOH$, and $CH_3CH_2CH_2OH$.

When a protein is introduced onto the reversed-phase column, the non-

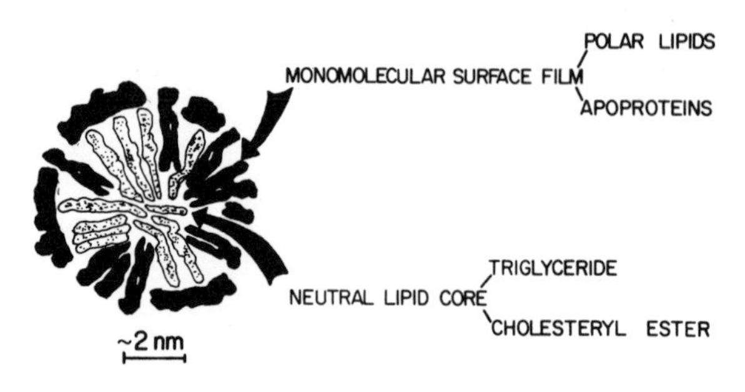

FIG. 2. A model for lipoprotein structure based on the interactions between apolipoproteins and lipid constituents. The surface monolayer is composed of phospholipids and apolipoproteins. The apoproteins contain helical regions which are amphipathic. The hydrophobic surface of the amphipathic helix interacts with the fatty acyl chains of phospholipids, and the hydrophilic surface is exposed to the aqueous environment of the polar head groups and the plasma. Adapted from Pownall *et al.*, (1981).

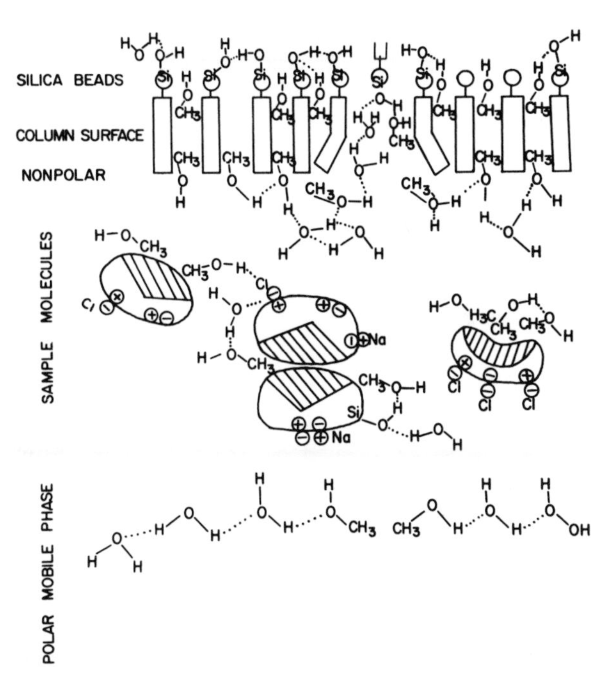

FIG. 3. A model for the interaction of a protein molecule and a reversed-phase column. The hydrophobic regions of the sample molecules displace solvent molecules from the adsorbed layer of the reversed phase. Adapted from Hancock and Sparrow (1981c).

polar regions of the molecule can displace some of the solvent molecules from the surface of the reversed phase. At this state the protein molecule will probably be adsorbed at the surface of the reversed-phase "bristle." It is unlikely that the protein will penetrate far into the reversed phase unless it contains very large nonpolar surface areas, and thus adsorption rather than partition is the preferred mechanism of interaction.

This proposed mechanism for protein separations is supported by the recent theoretical studies of Horváth *et al.* (*29*) and Horváth and Melander (*28*). In these studies, the hydrophobic effect in aqueous-organic systems (termed the solvophobic theory) was used to predict the retention of peptides on a nonpolar column. These authors found that the dominant interactions were between the mobile and stationary phases and between the mobile phase and the sample molecules. The driving force in both interactions was the shielding of a nonpolar region of either the column or sample molecule from the polar aqueous phase.

Based on this theory, one can explain the effect of an organic solvent gradient in displacing nonpolar samples from the column during a reversed-phase analysis. The solvent gradient reduces the surface tension of the water molecules and thus decreases the energy required to form a cavity of water molecules surrounding nonpolar molecules. With a sufficient decrease in surface tension, the sample molecule is no longer adsorbed to the reversed phase and is eluted in the mobile phase.

C. Prediction of Retention Times

The theory also predicts that the retention of a protein will be related to the nonpolar surface area of the sample. This prediction is supported by the observed retention time for the apolipoproteins C-I, C-II, and C-III on an alkylphenyl column in the presence of an acetonitrile gradient (Fig. 4). The observed elution order is consistent with the known polarity of these proteins, with the apolipoprotein of the greatest nonpolar surface areas being retained the longest (*15*). The isomers of the apolipoprotein C-III differ in the sialic acid content (0–2 residues) of the carbohydrate chain attached to Thr_{74}. One would expect that the polar sialic acid residue would decrease the nonpolar surface area of the protein molecule. The observed elution order of these isomers (Fig. 4) can be seen to be consistent with the sialic acid content.

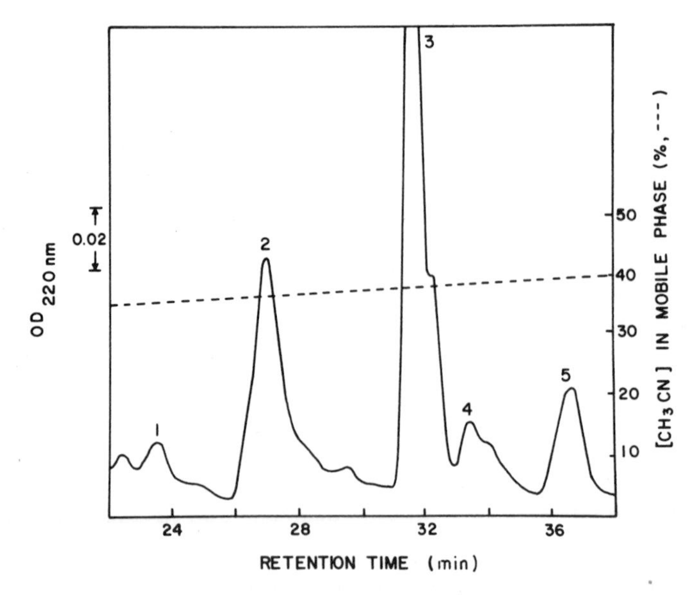

FIG. 4. The separation of a mixture of C-apolipoproteins on a μ-Bondapak alkylphenyl column with a mobile phase of 1% triethylammonium phosphate, pH 3.2, and an acetonitrile gradient. The flow rate was 1.5 ml/min, and the identity of the peaks was determined by amino acid analysis: apolipoproteins C-I (1), C-III_2 (2), C-III_1 (3), C-III_0 (4), C-II (5).

One can obtain some estimate of the surface polarity of a protein of known structure and hence its retention on a reversed-phase column by a consideration of the polarity of the side-chain functional groups. Such calculations have been carried out by a number of authors for amino acids (*23, 37*) and peptides (*35, 42*). The best correlation between the retention of a peptide (up to 15 residues) and its structure is obtained by the sum of the absolute fragmental hydrophobic constants (*41*) of the five most strongly hydrophobic residues (Trp, Phe, Leu, Ile, and Tyr).

Larger molecules such as proteins usually do not fit these predictions, probably because the molecules adopt an ordered three-dimensional structure in which many of the hydrophobic residues are buried within the structure and unavailable for interaction with the reversed phase. As might be expected from the proposed mechanism of separation, the retention of proteins on reversed-phase columns is not related to molecular weight of the sample, but rather the surface polarity of the molecule. Table I shows that there is a correlation of hydrophobicity (measured by mole % of strongly hydrophobic residues) with retention order for seven different proteins. It is unlikely that the retention of all proteins on a reversed-phase column can be correlated in this manner, because many protein structures have few nonpolar residues exposed to the aqueous environment. For example, although the major A and C apolipoproteins are eluted from a μ-Bondapak alkylphenyl column in an order which can be related to the proposed secondary structures, there is little correlation with the content of hydrophobic residues in each protein and the degree of interaction with the stationary phase. A similar lack of correlation be-

TABLE I

Correlation between Nonpolar Amino Acid Content and Retention Time for Protein Samples[a]

Protein	Retention time (min)[b]	Hydrophobic amino acids[c]	Protein	Retention time (min)[b]	Hydrophobic amino acids[c]
Lactalbumin	∞[d]	35.8	Cytochrome *c*	35.0	25.0
Elastase	∞[d]	32.5	Ribonuclease	27.5	21.8
Myoglobin	45.0	31.3	Neurotoxin 3	24.0	21.1
Lysozyme	37.5	26.4			

[a] Adapted from O'Hare and Nice (*43*).

[b] The protein standards were chromatographed on a Hypersil ODS column with a 0.1 *M* NaH_2PO_4–H_3PO_4 buffer and a gradient of 0–60% CH_3CN over 1 h (multistep).

[c] Sum of fragmental constants (*47*) for strongly hydrophobic residues, i.e., Trp, Phe, Leu, Ile, Tyr.

[d] Sample retained indefinitely.

tween relative retention values and lipophilicities was observed for various cytochromes (*65*) and α-lactalbumin (*39*).

The actual retention time of a protein will depend on the reversed phase, organic solvent, and nature of the mobile phase. It has been noted that anomalies occurred with closely related, small polypeptide hormones, even though a stable secondary structure was not expected for these molecules. Thus, Terebe *et al.* (*64*) have found that the elution order of a series of ACTH analogs is affected by the position of certain residues, for example, the interchange of residues 3 and 11 in the molecule (Ser and Lys) results in a significant difference in retention time. These authors suggest that a residue change in an endo position causes a frame shift in the molecule. This shift alters the steric relationships of the side chains, and can lead to a significant conformational change which results in different molecular hydrophobicities. It is also possible that subtle changes in conformations of protein samples in the complex milieu of the reversed-phase column will result in unpredictable changes in retention times. Thus, any rules for prediction of retention times should be treated only as a useful guide. In addition, most separations will be carried out on samples of unknown structure, and therefore no predictions can be made about the optimal separation conditions.

D. Resolving Power of Reverse-Phase HPLC

The precise dependence of retention time on the nonpolar surface and hence conformation of a protein sample can be used to explain the highly selective separations achieved between very similar samples by reversed-phase HPLC. An approximate limit for most other chromatographic separations is a difference of a single charged residue in a chain of 50 amino acids, or perhaps the difference of a single nonpolar residue in a total of 20 residues. The following examples will clearly demonstrate the potential of reversed-phase HPLC for the separation of closely related proteins. Porcine insulin is well separated from the monodesamido derivative (charge difference of −1) by isocratic elution from a LiChrosorb RP8 column (*7*). Rivier (*49*) separated bovine and porcine endorphin although the two samples differ only by a methyl group (at position 83 of the original β-lipotropin sequence). Two separate subunits of hemoglobin F (Gγ and Aγ) differ only by a methylene group but again are well separated by isocratic elution (*58*).

Another test of the resolving power of this new technique is to compare such separations with those achieved with conventional procedures. Glasel (*11*) found that neurophysins I and II, which were derived in an apparently chromatographically pure form by isoelectric focusing, were resolved into a number of components by a C_{18} reversed phase. Petrides *et al.* (*44*) reported that a sample of the β-chain of hemoglobin that had been purified by ion-exchange chromatography could be resolved into two peaks of similar amino acid analysis on a Supelcosil C_{18} column. Svoboda *et al.* (*62*) found that reversed-phase HPLC of partly purified somatomedin gave a sharp peak of activity, whereas isoelectric focusing gave incomplete purification. Hearn and Hancock (*25*) demonstrated that a sample of bovine thyroid-stimulating hormone which appeared to be homogeneous by ion-exchange chromatography was resolved into several peaks by chromatography on a μ-Bondapak C_{18} column. The observation of apparent microheterogeneity in protein samples, when chromatographed on reversed-phase columns, can be related to a number of causes. In a suboptimal chromatographic system a homogeneous sample can give rise to more than one peak due to causes such as restricted diffusion, steric exclusion effects, and aggregation or precipitation of the sample. Alternatively, the apparent microheterogeneity of the protein sample may in fact be real and be caused by structural differences, such as deamidation of side-chain amide groups or the presence of variable carbohydrate side chains.

II. PROPERTIES OF REVERSED-PHASE COLUMNS

A. Nature of the Column

Of the wide range of functional groups available as chemically bonded stationary phases, C_{18}, C_8, CN, or phenyl packings are normally used in protein separations. All column packings supposedly function by providing a nonpolar surface area for adsorption of the solute from the polar mobile phase. Figure 5 gives an example of the excellent separation that can be obtained for a number of protein standards with a C_{18} column. As will be seen in Table V, the C_{18} phase has been used in the majority of protein separations, although other reversed phases of higher polarities may give useful additional selectivities. Most C_{18} columns are strongly retentive for nonpolar compounds and allow excellent separations of such mixtures. With biological samples, however, it may not be possible to use the high amounts of organic solvent necessary to elute nonpolar samples from the column. For this reason, the C-8 and alkyl phenyl columns with lower surface loadings and hydrophobicities have proven useful in many protein separations. Figure 6 shows the successful elution of apolipoprotein A-I (243 residues) on a μ-Bondapak alkylphenyl and a Zorbax C8, but not from a Zorbax ODS (C_{18}) column, where the sample was retained

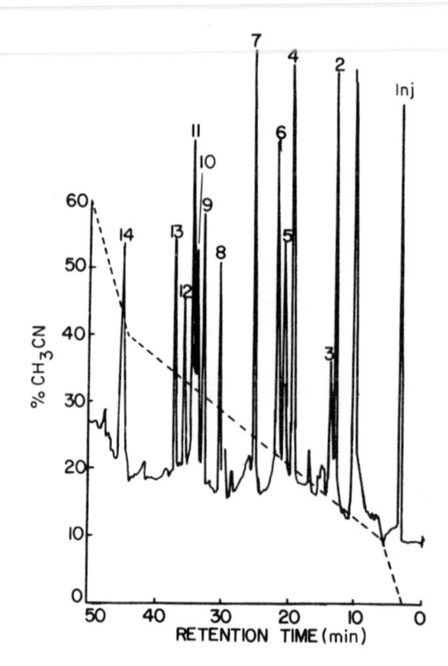

FIG. 5. The separation of polypeptide standards on Hypersil ODS with 0.1 *M* NaH_2PO_4–H_3PO_4, pH 2.1, as the mobile phase, at a flow rate of 1 ml/min. The peaks are as follows: 1, Trp; 2, lysine vasopressin; 3, arginine vasopressin; 4, oxytocin, 5, $ACTH_{1-24}$; 6, insulin A-chain; 7, bombesin; 8, substance P; 9, somatostatin; 10, insulin B-chain; 11, human calcitonin; 12, glucagon; 13, salmon calcitonin; 14, melittin. Adapted from Fig. 1 of O'Hare and Nice (1979).

indefinitely. Although gradient elution up to 80% acetonitrile was used in each case, the protein was not eluted from the Zorbax ODS column even at the later stages of the gradient, presumably because of precipitation and/or denaturation of the protein. In addition, the alkylphenyl column, which can give rise to specific π–π interactions with aromatic groups, can readily discriminate the aromatic amino acids, i.e., Phe, Tyr, His, and Trp, which are present in variable amounts in different proteins.

The following examples will be used to demonstrate that medium-polarity reversed phases have great potential for protein separations. In Fig. 7 the separation of the C-apolipoproteins is shown on a phenyl and on a C_{18} column. The C_{18} column gave almost no separation of the protein mixture, despite the use of a variety of mobile-phase conditions, whereas the phenyl column gave an excellent separation. Lewis *et al.* (*34*) found that the columns of intermediate polarity were effective in the separation of tyrosinase, collagen α_1, cytochrome *c*, and bovine serum albumin. In this study the diphenyl column was more retentive than the octyl or nitrile columns since 46% instead of 30–34% propanol was required in the mobile phase to elute the proteins. The diphenyl column also resolved Type I collagen into four components whereas the other columns only separated three peaks. Nice *et al.* (*40*) found that lactalbumin bound irreversibly to Hypersil ODS but could be eluted from Hypersil aminopropyl, although the chromatographic efficiencies were lower than normal. Subsequently, these authors demonstrated that Spherisorb C_6 gave the best recoveries of ovalbumin (*39*). Rivier (*49*) found that the nitrile column (Waters) was the best for separation of polypeptides and generally allowed the elution of samples with lower concentrations of organic modifiers in the mobile phase. Terabe *et al.* (*65*) found that amino but not C_{18} or CN reversed phases allowed separation of bovine and chicken cytochromes.

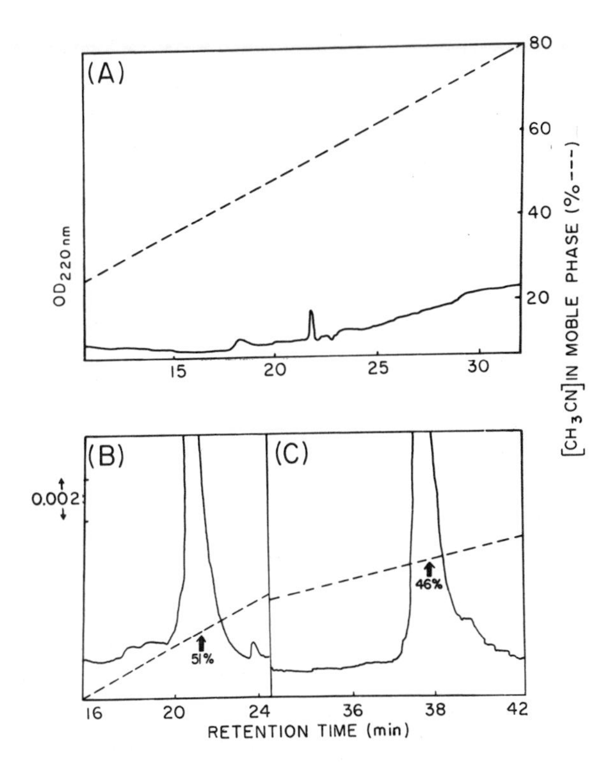

FIG. 6. The chromatography of apolipoprotein A-I on a Zorbax ODS (A), Zorbax C8 (B), and a μ-Bondapak alkylphenyl (C) column. In each case the mobile phase was 1% triethylammonium phosphate, pH 3.2, and an acetonitrile gradient was used. The flow rate was 1.5 ml/min. Each analysis was performed on 100 μg of protein, and the concentration of acetonitrile required to elute the sample is indicated in each panel. Adapted from Hancock and Sparrow (1981d).

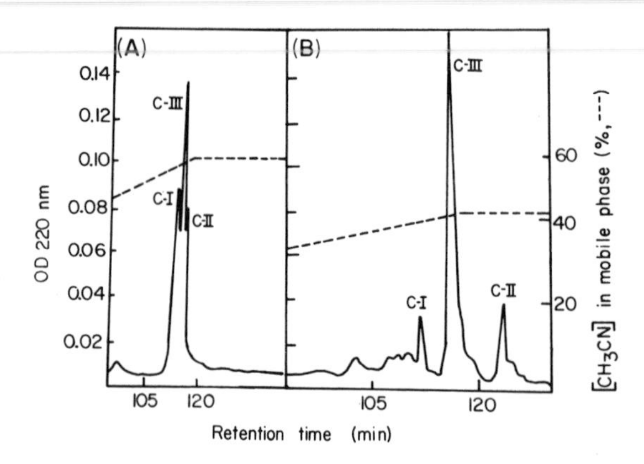

FIG. 7. The separation of a mixture of C-apolipoproteins on a C_{18} (A) and a phenyl (B) column (both Waters μ-Bondapak). All other chromatographic conditions are the same as described for Fig. 6.

However, protein samples exhibit great diversity and these observations about the suitability of medium-polarity reversed phases may not be applicable to all protein separations. For example, O'Hare and Nice (*43*) found that with a series of ACTH-related compounds, the precise nature of the column packing had little effect on the observed separations. Voskamp *et al.* (*68*) found that a nitrile column (Polygosil CN) gave a poor efficiency separation of secretin relative to C_{18} columns (Nucleosil and Lichrosorb). Biemond *et al.* (*3*) found that LiChrosorb RP18, which was the most hydrophobic stationary phase tested, gave the best selectivity in the chromatography of ACTH, β-endorphin, and insulin. Shelton *et al.* (*59*) found that a μ-Bondapak C_{18} column gave the best separation of globin chains.

Several other parameters, apart from the nature of the reversed phase, are important in the determination of the exact column best suited for a given separation. Such parameters as particle size, shape of particles, range of particle size, pore size, surface area of the silica particles per gram, amount of reversed phase per gram, the presence of polymerized surface coating, the use of a secondary silanization reagent, the stability of the reversed phase to aqueous conditions, and the column efficiency are all important details. In the following discussion, we will attempt to clarify the relevance of these column parameters to a protein separation.

B. Pore Size, Flow Rate, and Temperature Effects

As the molecular weight of the species to be separated increases, the pore size of the reversed-phase support becomes crucial, particularly since the nominal values listed for the parent silica must be reduced by at least 20–30 μm due to the bulk of the reversed phase (*13*). If the diffusion rate of the sample molecules in and out of the silica particles becomes slow, then broad, asymmetric peaks will be observed. In extreme cases, two separate peaks can be seen due to included and nonincluded material. For most commercial reversed phases with a pore size of 80–100 Å it is probable that protein molecules with MW > 20,000 cannot penetrate the pores leading to the interior of the silica particles and thus separation occurs on the surface of the particle. Such a mechanism would explain the lower capacity of reversed-phase columns for larger molecules. For example, the maximum loading of a Waters alkylphenyl column (pore size 80 Å) for Leu-Tyr-Gly-Phe-NH_2 was 800 μg and for apolipoprotein A-I (243 residues) was 200 μg. This example demonstrates, however, that larger proteins can be successfully chromatographed on such columns, as has been confirmed by other groups (see examples in Table II). It is possible, however, that many of these separations would be improved by the use of a reversed-phase support with larger pores (i.e., 200–500 Å), particularly if semipreparative separations were required. A danger with the use of even larger pores (>500 Å) is the possibility that such large pores can give rise to stagnant pools of solvent and poor diffusion rates (*66*). Lewis *et al.* (*34*) found that for increasing sample loads up to 200 μg there was no change in the chromatographic pattern obtained between low and high loadings for a reversed-phase column of either 100- or 500-Å pores. However, for a sample load greater than 200 μg, on the 100-Å column, some protein was not retained and peak width increased. The 500-Å column could be loaded with several milligrams of protein without overloading

TABLE II

Molecular Weights of Proteins Chromatographed Successfully by Reversed-Phase HPLC

Protein	MW	Protein	MW
Ferritin	450,000	Apolipoprotein A-II	17,400
Catalase	240,000	Lysozyme	14,300
Tyrosinase	128,000	Ribonuclease	13,700
Collagen α_1	95,000	Cytochrome *c*	11,700
Serum albumin	65,000	β-Lipotropin	11,600
Ovalbumin	43,000	Neurophysin	10,000
Thyroid-stimulating hormone	30,000	Apolipoprotein C-III	9600
		Apolipoprotein C-II	8800
Apolipoprotein A-I	28,000	Acyl carrier protein	8800
Chymotrypsinogen A	25,000	Neurotoxin 3	7800
Growth hormone	19,000	Apolipoprotein C-I	6600
Interferon	18,000	Insulin	5400

TABLE III

Peak Width Obtained for Chromatography of Proteins on Octyl-Silica of Different Pore Sizes[a,b]

Protein	Molecular weight	Peak width (mm)	
		100 Å	500 Å
Collagen α_1	95,000	13.0	9.5
Bovine serum albumin	68,000	9.0	6.0
Cytochrome *c*	12,500	5.5	6.0

[a] This table was adapted from the study by Lewis *et al.* (*34*).
[b] A 10-μg sample of each protein was chromatographed on the C_8 column at a flow rate of 30 ml/h.

effects. Table III lists the proteins studied and shows that even at low loadings proteins with MW > 20,000 show broader peaks on the 100-Å than on the 500-Å column. A small protein such as cytochrome *c*, however, chromatographs with a slightly broader peak on the 500-Å column. In this case the pore size of the 100-Å column is not limiting, and the observed difference may be due to some stagnant pools of solvent with the 500-Å support. Nice *et al.* (*39*) found that large-pore-size (300 Å) supports gave improved efficiencies in separation of protein standards.

Clearly, these studies demonstrated that it is desirable to chromatograph a high-MW protein sample on a wide-pore (200–500 Å) silica coated with a medium-polarity reversed phase. Unfortunately, manufactured columns generally do not meet these criteria, and thus the researcher must either exercise caution in the selection of a commercial column or prepare his own packing material.

Lower flow rates are important in the chromatography of proteins and can partly compensate for the lower diffusion rates of these macromolecules in the chromatographic column. Table V shows that flow rates of 30–90 ml/h are commonly used, although the lower range is best for proteins with MW > 10,000.

Another parameter which can affect diffusion rates of solutes is temperature. Not surprisingly, protein separations are usually carried out at room temperature, although O'Hare and Nice (*43*) found that $ACTH_{1-39}$ was chromatographed at a greater efficiency at 70°C than at 40°C (26,800 versus 15,000 theoretical plates/m at a flow rate of 1 ml/min). However, Rivier (*49*) found for other protein standards and Terabe *et al.* (*65*) for cytochromes that the opposite was true, and increased temperature decreased separation efficiency. In view of the limited stability of most proteins at higher temperature, it is likely that most separations will still be carried out at ambient or lower temperatures.

C. Silanol Content and Loading of Reversed Phase

The silanol content of a reversed-phase column is a very important parameter which can dramatically affect the selectivity of the column for a series of compounds. In general, the higher the silanol content the more polar the column and the shorter the retention of nonpolar compounds. Silanol groups are always present in a reversed-phase packing either due to incomplete reaction of the organochlorosilane used in introduction of the reversed phase, or to hydrolysis of unreacted chlorosilyl groups left after introduction of the reversed phase. These silanol groups can be reduced by a process known as "capping" in which the support is derivatized again with a small reactive silane such as trimethylchlorosilane. The silanol groups present in a reversed-phase packing would be expected to increase the degree of hydration of the support. Any increase in hydration should increase mass transfer rates of polar solutes and thus be of benefit in the analysis of proteins. For example, a C_{18} column with a high silanol content has been used for the separation of a complex mixture of peptide fragments and for apolipoproteins (*18*). An example of the highly resolutive peptide map which can be obtained with the radially compressed column is shown in Fig. 8. It should not be forgotten, however, that silanol groups, while increasing the polarity of the column, can lead to poor peak shapes and recoveries. The success of these examples can,

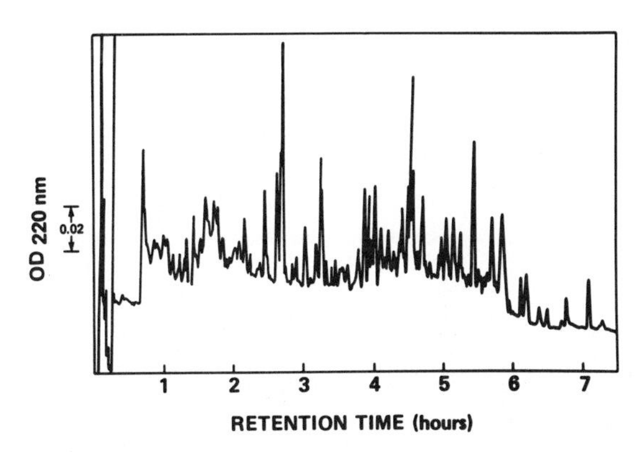

FIG. 8. The peptide map for the tryptic cleavage of apolipoprotein B which was obtained on a Waters radially compressed analytical system with a mobile phase of 1% triethylammonium phosphate, pH 3.2, and a gradient of acetonitrile. The flow rate was 1 ml/min. Adapted from Hancock and Sparrow (1981d).

therefore, be attributed to the blocking of silanol group interactions with the sample (by steric factors and by high salt concentrations in the mobile phase), while taking advantage of the polarity effects of this group.

Due to the large molecular dimensions of proteins, these molecules probably will be unable to interact directly with most silanol groups present on the support. Therefore, one could expect a difference in peak shape and recoveries between peptides and proteins on supports which have a relatively high silanol content, for example, the C_{18} column produced by Waters Associates for use in their radially compressed analytical system. In fact, we have found that this column can give problems with small basic peptides but is particularly useful for the separation of larger peptides and of proteins (*20*).

Another example of the effect of silanol groups on the chromatography of a protein sample is seen in Fig. 9. A sample of apolipoprotein A-I was analyzed both on a Spherisorb ODS (high loading of organic phase) and on a C_{18} reversed phase with many free silanol groups (Waters, radially compressed analytical column). The facile elution of the protein from the C_{18} column with the high silanol content can again be attributed to the polar effect of these groups and/or the lower surface coverage of the silica particles with the reversed phase.

Petrides *et al.* (*44*) used Supelcosil C_{18}, which has high organic loadings,

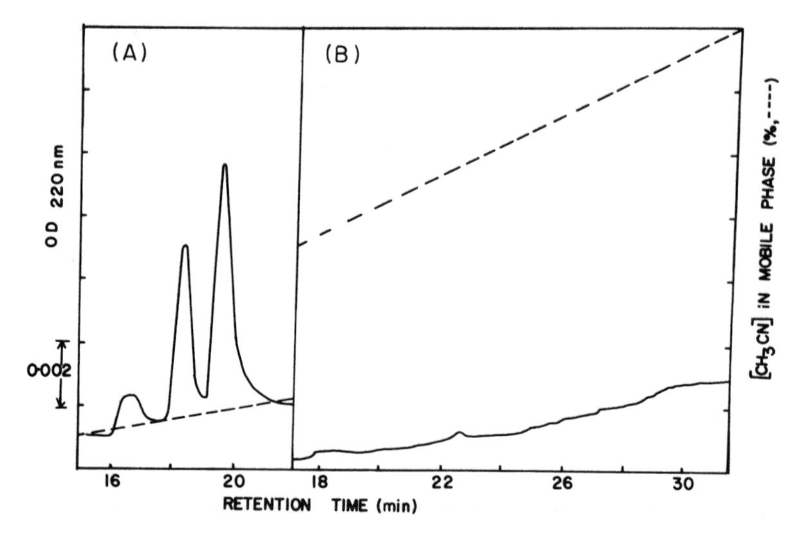

FIG. 9. The chromatography of apolipoprotein A-I on a Waters C_{18} radially compressed column (A) and on Spherisorb ODS (B). In each case the mobile phase was 1% triethylammonium phosphate and the flow rate was 1.5 ml/min. Both analyses were made on 100 μg of protein. Adapted from Hancock and Sparrow (1981d).

for the separation of hemoglobin chains. This support gave better resolution of the samples than more polar columns, but still did not give an optimal separation as broad peaks were observed. This observation was attributed to slow mass transfer between mobile and stationary phases. Shelton *et al.* (*58*) found that a μ-Bondapak C_{18} column gave better separation of globin chains than a Zorbax ODS column, presumably due to the lower C_{18} loading in the former packing. It should be noted that the Bondapak packing is end-capped, unlike the Zorbax packing.

Another approach to the preparation of a column of low hydrophobicity is to use an organosilane with a short C-chain, e.g., trimethylchlorosilane. However, Nice *et al.* (*40*) found that such a column (Hypersil SAS) deteriorated rapidly over 24 h when used with an aqueous mobile phase. The "aged" column exhibited loss of retention for polypeptides, and analysis showed an increase in the silanol content of the packing.

These examples illustrate that the silanol content of a reversed-phase packing is important and should be compatible with the protein under study. Cationic proteins may well exhibit low recoveries on packing materials with a high silanol content, whereas other proteins (particularly very hydrophobic ones) may chromatograph well on these more polar supports. The next section on the properties of the mobile phase will describe methods for minimizing the undesirable properties of silanol groups.

III. PROPERTIES OF MOBILE PHASES

A. Function of Organic Modifier

At first sight, it would appear that many biological molecules would be too polar to interact significantly with a nonpolar stationary phase, and thus would be unretarded by the column. The enormous surface area of a microparticulate support combined with at least a small amount of nonpolar surface area for most proteins does allow, however, considerable interaction between the sample and the long-chain alkyl group of a μ-Bondapak C_{18} column, or the phenyl group of a μ-Bondapak alkylphenyl column. In fact many protein samples cannot be eluted from the reversed-phase column. Such samples interact strongly with the nonpolar reversed phase and/or the unreacted silanol groups on the column. To decrease this interaction and thus allow the sample to elute at a convenient retention time, it is necessary to add organic solvent and salts to the mobile phase. In an earlier section the retentions of solutes on a hydrocarbonaceous surface was related to the hydrophobic effect in mixed aqueous–organic solvents. In such a system, the effect of an organic solvent was to lower the surface tension of the aqueous phase and thus facilitate the transfer of the sample from the stationary to the mobile phase. The lower the polarity of the organic solvent the greater the decrease in retention time observed for a protein at a certain amount of the solvent in the mobile phase (*43*). The use of very nonpolar solvents is generally limited by lack of solubility in water, although some may prove useful in ternary solvent mixtures. In general, the solvent chosen should give the best separation at the lowest possible concentration of organic modifier so that precipitation or denaturation of the sample is minimized.

Many protein isolations also involve the use of organic solvents at one or more purification steps, and thus one can often use this solvent in HPLC to maximize the recovery of protein samples. For example, apolipoproteins can be solubilized in isopropanol without loss of activity. For this reason Hancock and Sparrow (*20*) made extensive use of this solvent in the separation of C-apolipoproteins by reversed-phase HPLC. Glasel (*11*) found that neurophysins had a high solubility in aqueous methanol and used this solvent mixture in HPLC studies.

In addition, the use of different solvents can alter the elution order of protein mixtures. O'Hare and Nice (*43*) found that the elution order of insulin and $ACTH_{1-39}$ from a Hypersil ODS column could be inverted by a change from methanol to acetonitrile as the organic modifier. A recent development in liquid chromatography is the use of ternary solvent mixtures, and although such mixtures have had little use in protein separations, the addition of small amounts of solvents such as tetrahydrofuran ($\epsilon^\circ = 0.45$) or dioxane ($\epsilon^\circ = 0.56$) may improve the separation of nonpolar components when acetonitrile ($\epsilon^\circ = 0.65$) is used as the organic modifier. Alternatively more polar solvents such as methanol ($\epsilon^\circ = 0.95$), ethylene glycol ($\epsilon^\circ = 1.11$), and formamide ($\epsilon^\circ > 1.2$) may be used to improve separation of polar components. Organic components which have very limited solubility in water, such as butanol, hexanol, and dodecylamine, can also be introduced into an organic modifier such as acetonitrile so that the mobile phase is not limited to water-soluble solvents. Gazdag and Szepesi (*9*) found that isopropanol increased the separation of protein standards.

B. Effect of Mobile Phase on Properties of Stationary Phase

The silanol group (Si-OH), present at least at low concentrations in all reversed-phase columns, is capable of adsorbing a layer of a polar solvent such as methanol or water. A column conditioning procedure based on this adsorption is an essential step in normal-phase chromatography on silica columns. This process is probably also important in reversed-phase chromatography where adsorption of water or an alcohol to surface silanol groups may enhance partitioning of proteins from an aqueous mobile phase to the stationary phase.

If the mobile phase contains a less polar solvent such as butanol, then one could expect interaction between the nonpolar reversed phase and nonpolar regions of the solvent molecules. Such a solvation process may be important in determining the conformation of the nonpolar bristles of a reversed phase in the presence of an aqueous mobile phase. Figure 10 shows an artist's visualization of the solvation processes and illustrates how the surface chemistry of the reversed phase can be significantly affected by solvent interactions. Such a concept is consistent with the data of Scott and Kucera (*55*), who demonstrated that a reversed phase can adsorb a monomolecular layer of either polar or nonpolar solvent molecules. The scheme is also consistent with the observation that many pro-

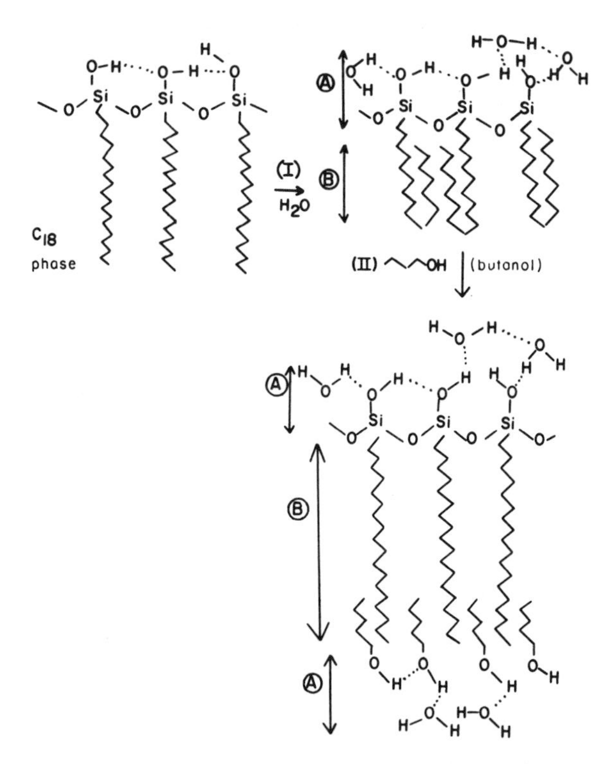

FIG. 10. An artist's visualization of a C_{18} reversed phase in the presence of an aqueous mobile phase (I) and in the presence of butanol (II). The polar and nonpolar regions are denoted by the letters A and B, respectively. The possible conformations of the alkyl chain are represented by the sawtooth figures. Adapted from Hancock and Sparrow (1981d).

tein separations are better carried out under gradient than isocratic conditions, where the column is exposed to both extremes of solvent polarity (*19*). Monch and Dehnen (*38*) proposed that 2-methoxyethanol could act as a surface-active agent in a similar manner to butanol. It was proposed that, in the presence of such an organic modifier, proteins would be less strongly bound to the column and thus eluted more readily.

C. Effect on Retention Times

It has been noted in a number of separations that the retention time observed for a particular protein is very sensitive to the concentration of organic modifier in the mobile phase. As is shown in Fig. 11, the capacity factor (K', a measure of retention time) is much more sensitive for large peptide and protein samples, e.g., $ACTH_{1-24}$, than for small peptides, e.g., $ACTH_{5-10}$ and Phe-Phe. Shelton *et al.* (*58*) found that the separation of globin chains was very sensitive to changes in the level of acetonitrile. Hancock and Sparrow (*20*) found that five different apolipoproteins (C-I, C-II, C-III_2, C-III_1, C-III_0) could be eluted by a very shallow gradient of acetonitrile (over 5% v/v). It was necessary, however, to use a gradient to achieve this separation. Isocratic elution conditions using the acetonitrile concentration at the start of the gradient gave good separation of the early-eluting peaks, but infinite retention of the later-eluting apolipoprotein C-II. Subsequent elution of the sample with a higher acetonitrile concentration gave poor recoveries of the protein (<40%). Alternatively, isocratic elution using the acetonitrile conditions corresponding to the end of the gradient gave rapid elution of all components of the mixture and little separation. O'Hare and Nice (*43*) made a similar observation for the

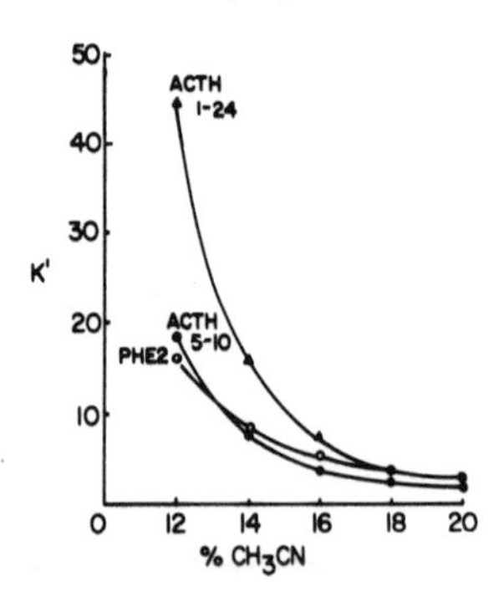

FIG. 11. The effect of acetonitrile concentration on the retention times of polypeptides using a Hypersil ODS column with a mobile phase of 0.1 *M* NaH_2PO_4–H_3PO_4, pH 2.1, and acetonitrile at the concentration shown in the figure. The flow rate was 1 ml/min. Adapted from O'Hare and Nice (1979).

chromatography of β-lipotropin, ACTH, and β-endorphin, where a very narrow range of acetonitrile (3–4%) could be used to separate the mixture. Acetonitrile above and below these limits gave either no retention or irreversible binding. The sensitivity of protein elution to the concentration of organic modifier is an advantage in gradient elution, however, because it can lead to sharp peaks and reproducible elution times. Lewis *et al.* (*34*) found that bovine serum albumin could be eluted from a reversed-phase column with 34% *n*-propanol within three column volumes. If the mobile phase contained 32% *n*-propanol, the protein was not eluted even after 10 column volumes of mobile phase. The sample could, however, be satisfactorily eluted by returning the mobile phase to 34% *n*-propanol. Petrides *et al.* (*44*) found that lower concentrations of propanol in the mobile phase gave optimal separation of globin chains but at lower recoveries. These results suggest that different proteins will respond in a variable manner to adsorption onto a reversed-phase column and delayed elution with an organic modifier. It seems to be a general rule, however, that a protein separation is very sensitive to slight changes in solvent composition.

D. Organic Solvent Gradients

The previous section suggested that for high-efficiency separations it is necessary that sample molecules do not interact too strongly with the reversed phase by either adsorption or silanol interactions. This requirement probably explains why many protein separations are better carried out with shallow gradients rather than isocratic separations. The gradual increase in organic modifier concentration would serve to continually displace the protein molecules from adsorption sites before irreversible multipoint binding occurs. Figure 12 shows the use of gradient analysis to establish the separation conditions for a complex mixture of C-apolipoproteins. The advantage of a full gradient analysis is that all components in the mixture will be eluted regardless of the polarity of the sample. The disadvantage of a quick scanning of the sample mixture is that some components may be partially precipitated on the column. Once the correct gradient shape and level of organic modifier have been established (Fig. 12d), then the time of the gradient analysis should be increased. In Fig. 4 the separation conditions actually used for this protein mixture are shown. The longer and shallower gradient allowed better resolution of the components of the mixture, good recoveries (87–95%), and minimal precipitation of the sample on the column. The reverse gradient should always be monitored as it returns to the initial conditions since it can show if any of the sample was precipitated during the original gradient analysis. If precipitation does occur, then some of the sample

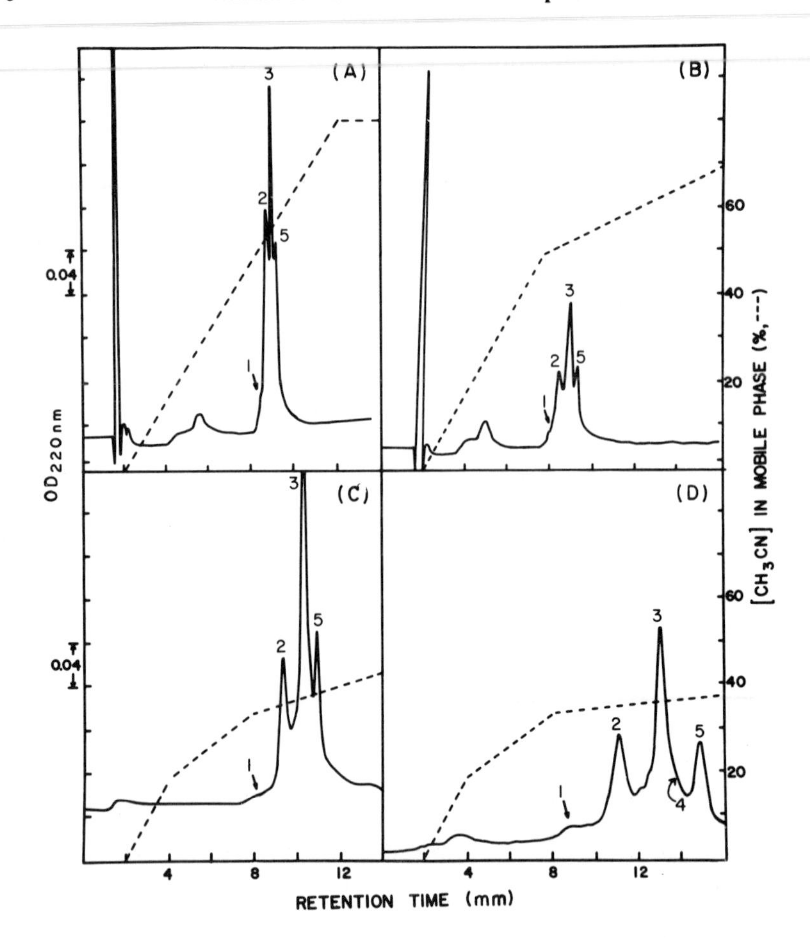

FIG. 12. The separation of a mixture of C-apolipoproteins (VLDL) on a Waters μ-Bondapak alkylphenyl column with a mobile phase of 1% triethylammonium phosphate, pH 3.2, and acetonitrile as the organic modifier using several gradient slopes. The flow rate was 1.5 ml/min. The different proteins were identified by amino acid analysis and pure standards as follows: 1, apolipoprotein C-I; 2–4, apolipoprotein C-III with 2, 1, and 0 mol of sialic acid in the carbohydrate side chains, respectively; 5, apolipoprotein C-II. Adapted from Hancock and Sparrow (1981c).

should dissolve at the same mobile-phase composition in the reverse as in the initial gradient analysis. If a slow gradient analysis is used as well as pure solvents, then no major optical density peaks should be observed in the reverse gradient apart from the optical density difference of the solvents. With the apolipoprotein mixture the elution of protein material during the reverse gradient was not observed. A blank gradient run immediately after this analysis showed that the contamination between analyses was less than 0.6%.

E. Identification of Optical Density Peaks

The initial gradient analysis (such as shown in Fig. 12) does not guarantee that the sample will be eluted from the column or be separated from other components of the mixture. For example, the sample may be bound irreversibly to the column, precipitated by the organic solvent, obscured by the solvent peaks in the breakthrough volume, or obscured by peaks from buffer or solvent impurities present in the gradient run. For these reasons it is essential that the initial gradient analysis be followed by some identification procedure to confirm that the sample of interest has in fact been eluted satisfactorily. Some widely used procedures are the following:

(1) Retention time analysis
(2) Co-chromatography with reference compounds
(3) Adsorbance ratios, e.g., the monitoring of the effluent at 215 and 280 nm
(4) Fluorescent monitoring of effluent (either specific for a fluorogenic residue, e.g., Trp, or the presence of a free amino group if a fluorescent derivative is introduced)
(5) Stopped-flow UV and fluorescent spectra
(6) Radioactive detection
(7) Enzymatic or chemical peak-shift technique (use of a specific chemical or enzymic reaction to remove selected components of the mixture)
(8) Collection of the peak and direct identification by a technique such as amino acid analysis or mass spectroscopy

IV. USE OF ION-PAIRING REAGENTS

A. Mechanism of Action

The addition of millimolar concentrations of most salts to the mobile phase has the effect of decreasing the retention time of a protein sample and improving peak shapes. This effect can be attributed to suppression of the interaction between polar groups on the solute and silanol groups. Other effects such as ion pairing can also operate and these will be described here. Higher salt concentrations may result in increased retention times due to "salting-in" of the protein into the reversed phase, but this usually occurs at much higher salt concentrations.

In addition to this general electrolyte effect, various salts, when added to the mobile phase, can give rise to significant differences in peak shapes and retention times for a given sample. Such differences can give important additional flexibility to the chromatographic system. In an attempt to explain such effects, the concept of ion pairing was introduced originally by Schill (*53*) and subsequently applied to the HPLC analysis of a variety of materials (see Ref. *22* for an extensive review).

The concept of ion pairing refers to the ionic association between ions added to the mobile phase and oppositely charged ionized groups of the sample molecule, i.e.,

$$—\overset{+}{N}H_3 + X^- \rightarrow —\overset{+}{N}H_3X^- \quad \text{or} \quad —COO^- + Y^+ \rightarrow CO\overset{-}{O}\overset{+}{Y}$$

Depending on the polarity of the counterion X or Y, such an association can lead to a complex of either increased or decreased polarity. Such flexibility is important since different protein samples exhibit a wide range in polarity. An extremely nonpolar protein will be retained too long on a C_{18} column, whereas a very polar protein will not be sufficiently separated from the solvent peak even in the absence of organic modifier. With simple organic molecules, steep solvent gradients can be used for extremely nonpolar samples, but high organic solvent levels often precipitate or denature proteins and thus cannot be used. With the use of a nonpolar ion-pairing reagent a very polar sample can be retarded, whereas the use of a polar ion pair can lead to elution of nonpolar proteins at reasonable levels of organic solvent.

In an attempt to rationalize the large variety of buffers used to date in protein HPLC separations, it is necessary to briefly return to the separation mechanism based on the hydrophobic effect. This mechanism related the retention of a sample on a reversed-phase column to the nonpolar surface area of a sample molecule. A possible explanation of the dramatic effect of ion-pairing reagents on the retention of proteins on reversed-phase columns can then be based on the modification of surface polarity of the protein molecule on association with suitable counterions. In the absence of salts dissolved in the mobile phase, the peptide or protein sample probably has some counterions associated with the sample. Alternatively, the basic side chains of the protein may be neutralized by a salt bridge with an acidic residue which is adjacent in the three-dimensional structure. In either case, the contribution of the ammonium group to the surface polarity is relatively small. Figure 13 shows the result of association of the amine cation with a highly polar anion such as phosphate, which has a substantial sphere of hydration. In this case, the nonpolar area of the

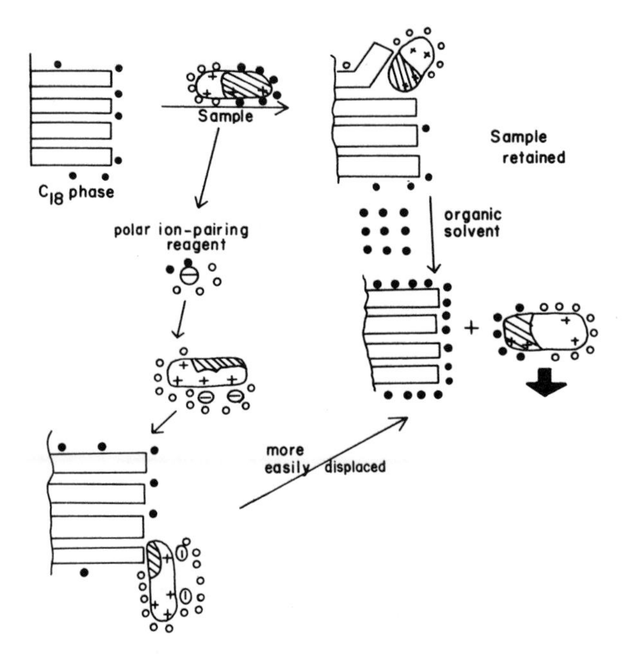

FIG. 13. A possible mechanism for the effect of ion-pairing reagents on the retention of a peptide or protein on a nonpolar reversed phase. The polar ion-pairing reagent reduces the nonpolar area of the protein, allowing it to be more easily displaced by organic modifier. ○, water molecule; ●, organic solvent molecule; ▨ nonpolar region; ☐, polar region. Adapted from Hancock and Sparrow (1981c).

protein molecule which is available for interaction with the reversed phase, is substantially reduced.

This pictorial representation of a possible mechanism of action of a polar ion-pairing reagent has allowed prediction of what anions will be useful in decreasing the retention of a given protein. Alternatively, a nonpolar anion such as hexane sulfonate could be expected to associate with an ammonium ion present in the sample, with an increase in the nonpolar surface area and hence retention time. For example, apolipoprotein C-III, which is readily eluted by an organic solvent gradient from a C_{18} column in the presence of phosphate anions, is retained indefinitely if butanesulfonate is used as the counterion.

B. Examples of Useful Ion-Pairing Reagents

In Table IV, a list of ion-pairing reagents is shown and illustrates the variety of combinations which can be achieved. In addition to anionic

TABLE IV

Anionic and Cationic Counterions Useful in Reversed-Phase HPLC

Anionic counterions	
Chloride	C_1–C_{16} sulfonate
Bromide	Toluene sulfonate
Iodide	Naphthalene sulfonate
Perchlorate	Camphor sulfonate
Phosphate	C_1–C_{16} sulfate
Acetate	Butylphosphate
Propionate	Citrate
Picrate	Tartrate
	Trifluoroacetate

Cationic counterions
(a) R_4NH^+, where R = C_1 – C_7
(b) R_3NH^+, where R = C_2, C_8, or C_{10}
(c) $R_2NH_2^+$, where R = C_1 or C_2
(d) where R = C_1–C_{12} or 2-hydroxyethyl-
(e) Inorganic ions, e.g., Na^+, K^+, Li^+, Cs^+, Mg^{2+}, Ca^{2+}

reagents, ion pairing can presumably be achieved with carboxylate ions of proteins and suitable cationic reagents. It must be remembered that when a salt is added to the mobile phase, both the anionic and cationic constituents can interact with the sample molecules. Although the effect of different cations was shown to be small for a number of peptides (*14*), the extra selectivity obtained with these ions may well be useful in the analysis of proteins rich in aspartic and glutamic acid residues.

The great success of a variety of amine phosphates such as ammonium phosphate, ethanolamine phosphate, cyclohexylammonium phosphate (*14*), or triethylammonium phosphate (*49*) as ionic modifiers can be attributed to a variety of effects. The relatively high salt concentrations used (typically 0.1–0.2 *M*) combined with the use of both polar anions and cations (which can ion-pair with ionic groups present on the sample) act to prevent unfavorable interactions between the sample and silanol groups present in the column packing. In addition, the use of amine cations appears to be particularly effective in deactivating silanol groups, presumably by a mixture of hydrogen bonding and ionic interactions. One should not ignore the effect of any added salt on the structure of the water component of the mobile phase and on the tertiary structure of the protein, but it is unclear at this time what relationship such changes would have to the resulting separation.

C. Stationary-Phase Modification

When hydrophobic ion-pairing reagents, which have a significant nonpolar region, are added to the mobile phase an additional event can occur, namely modification of the chromatographic support. These reagents have the potential to modify the nonpolar stationary phase by the formation of a surface layer of different polarity and/or charge. To interact significantly with the reversed phase, the modifier must contain a substantial nonpolar region. For example, hexane sulfonate is thought to act predominantly by ion pairing with the sample molecules (*55*) and not by stationary-phase modification (see Fig. 14A for a representation of this situation). Conversely, dodecyl sulfate readily coats the column and forms a dynamic cation exchanger (*37*), as well as ion-pairing with the sample (Fig. 14B). In a similar manner, dodecylammonium salts can be used to form a dynamic anion exchanger (Fig. 14C) (*14*). In addition, some ion pairing with carboxylate groups of the sample also probably occurs if the pH of the mobile phase is greater than 3.5. The low solubility of dodecylammonium salts in water has limited the use of this modifier to mobile phases with a high level of organic solvent. With an aqueous mobile phase, one can substitute the tetrabutylammonium salts which have good water solubility (*14*). Due to the instability of silica-based supports at extreme pH values, the normal pH range used in HPLC is 2.5–7. Almost all protein samples will be cationic at these pH values and therefore anionic reagents such as sodium

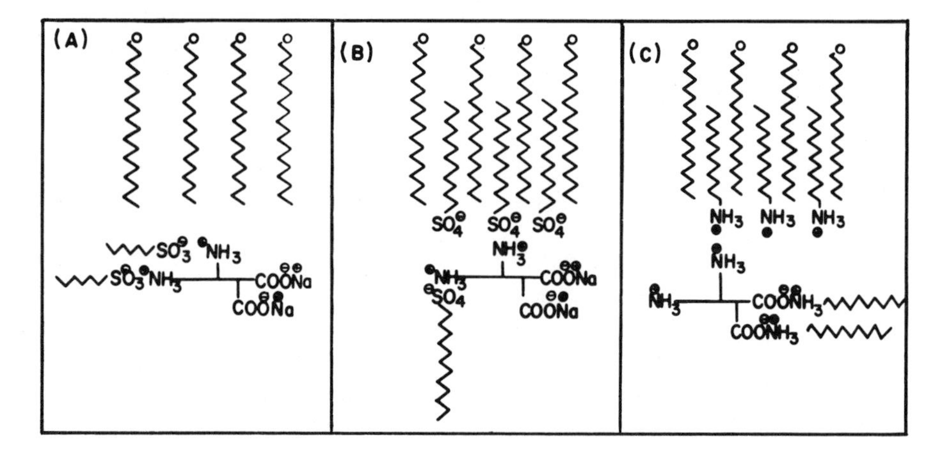

FIG. 14. An artist's visualization of possible interactions between the nonpolar reversed phase (shown as the sawtooth projections), the sample molecules, and the modifier: A and B, anionic modifiers of medium and low polarity, respectively; C, the effect of a low-polarity cationic modifier. Adapted from Hancock and Sparrow (1981c).

dodecyl sulfate (SDS) should prove to be more useful in protein analysis. In fact Rivier (*49*) found that the use of SDS in the mobile phase dramatically increased recoveries of protein samples.

D. Examples of Protein Separations Achieved with Ion-Pairing Reagents

It can be seen from the preceding discussion that an ionic material dissolved in the mobile phase can affect a number of parameters in a chromatographic system. Although some approximate predictions can be made about the effect of ion-pairing reagents on the retention times observed for a given protein, practical results are clearly of importance in evaluating these reagents. The following discussion will, therefore, list some observations made in the chromatography of different protein samples. Hearn and Hancock (*25*) found that low pH values (2.1–3) and relatively high concentrations of phosphate (0.1 *M*) were suitable for the chromatography of proteins such as proinsulin, thyroid-stimulating hormone, growth hormone, and acyl carrier protein. Conversely proteins such as aldehyde dehydrogenase and apolipoproteins C-I, C-II, C-III, and A-I could not be recovered from a reversed-phase column with such a mobile phase. As is described in Table V, aldehyde dehydrogenase could be successfully chromatographed if the mobile phase included dodecylammonium phosphate. The apolipoproteins were separated with a mobile phase which contained 0.17 *M* triethylammonium phosphate, pH 3.2.

Noskamp *et al.* (*68*) found that for the chromatography of secretin, 0.1% phosphoric acid gave very broad peaks, but that mobile phases which contained either KH_2PO_4 or 0.1% trifluoroacetic acid gave usable results. Biemond *et al.* (*3*) found that mobile phases which contained phosphoric acid gave better results than those with perchloric acid for the separation of ACTH, β-endorphin, and insulin. The best mobile phase found by these authors was 0.03 *M* tetramethylammonium phosphate, which resulted in a significant improvement in peak shapes but did not effect retention times. Lambert and Roos (*33*) found that nonpolar ion-pairing reagents, e.g., hexane sulfonate or tetrabutylammonium ions, gave better recoveries and resolution of calcitonins than did phosphates. Terabe *et al.* (*65*), however, found that alkylsulfonates had no effect on the chromatography of cytochromes and used a mobile phase which contained phosphates. Figure 15 shows the excellent separation which is achieved between bovine and porcine insulin with this mobile phase. Monch and Dehen (*38*), Nice and O'Hare (*41*), O'Hare and Nice (*43*), and Rivier (*49*) have confirmed our observations that for many protein separa-

TABLE V

Examples of Proteins Purified by Reversed-Phase HPLC

Sample	Column[a]	Mobile phase[b]	Reference
		Hormones	
1. Insulin	L-C_{18}	0.2 *M* $(NH_4)_2SO_4$, pH 3.5, 24.5% CH_3CN	*1*
2. Insulin	N-C_{18}	0.01 *M* $(NH_4)_2SO_4$, pH 2.2, CH_3OH : CH_3CN : H_2O, 5 : 1 : 4, 1 ml/min	*63*
3. ACTH, β-endorphin, glucagon, and insulin	L-C_{18}	0.5 *M* TMAP, pH 3, 26–41% CH_3OH, 1 h, 1 ml/min	*3*
4. ACTH precursor	μ-C_{18}	0.02 *M* TEAP, pH 5–90% CH_3CN, 90 min, 1 ml/min	*6*
5. Novo-insulin	μ-C_{18}	1% TMAP, pH 3, 40% CH_3CN, 1 ml/min	*25*
6. Insulins	μ-C_{18}	0.1% TEAP, pH 3.5, 26% CH_3CN, 2 ml/min	*25*
7. ACTH, insulins	N-C_{18}	5 m*M* tartrate, pH 3, 5 m*M* C_4SO_3Na and 0.1 *M* Na_2SO_4, 29% CH_3CN, 1 ml/min	*64*
8. ACTH, LPH	μ-CN	0.02 *M* TEAP, pH 3, 20–60% CH_3CN, 1 h, 0.5 ml/min	*57*
9. β-Endorphin	μ-C_{18}	10 m*M* formic acid, 50% CH_3CN, 1 ml/min	*10*
10. β-Endorphin	μ-C_{18}	0.05 *M* TMAP, pH 3, 31.5–80% CH_3CN, 1 ml/min	*67*
11. α-Neo-endorphin	μ-C_{18}	10 m*M* NH_4 formate, pH 4, 10–50% CH_3CN, 40 min, 2 ml/min	*30*
12. β-Lipotropin, β-endorphin	L-C_{18}	1 *M* Pyr-AcOH, 0–40% PrOH, 2 h, 0.5 ml/min	*51*
13. β-Lipotropin, ACTH, and β-endorphin	H-C_{18}	0.2 *M* P, pH 2.1, 0–60% CH_3CN, 1 h, 1 ml/min	*41*
14. Proinsulin	μ-C_{18}	0.1% P, 10–75% CH_3CN, 30 min, 2 ml/min	*25*
15. γ_3-Melanotropin	μ-C_{18}	0.1% TFA or 0.13% HFBA, 20–30% CH_3CN, 30 min	*4*
16. Corticotropin- and melanotropin-related peptides	μ-C_{18}	0.13% HFBA, 20–45.6% CH_3CN, 60 min, 1.5 ml/min	*5*
17. Corticotropin	μ-C_{18}	0.1% TFA, 20–40% CH_3CN, 1 h, 1.5 ml/min	*2*
18. Parathyroid hormone	H-C_{18}	0.155 *M* NaCl, pH 2.1, 0–50% CH_3CN, 1 h, 1 ml/min	*69*
19. Thyroid-stimulating hormone	μ-C_{18}	10 m*M* P, pH 2.1, 0–78% CH_3OH, 1 h, 2.5 ml/min	*23*
20. Growth hormone	μ-C_{18}	10 m*M* P, pH 2.1, 0–78% CH_3OH, 1 h, 2.5 ml/min	*23*
21. Growth hormone		8% Pyr, 2.5% AcOH, 0–100% CH_3CN : *i*PrOH, 1 : 1, 4 ml/min	*36*

(*Continued*)

TABLE V (*Continued*)

Sample	Column[a]	Mobile phase[b]	Reference
		Protein standards	
22. Tyrosinase, collagen α_1, cytochrome *c*, BSA	C-8, CN, diphenyl	0.4 *M* Pyr-formate, pH 4, PrOH, 0–40%, 2 h, 0.5 ml/min	*34*
23. Insulin, cytochrome *c*, BSA, catalase, ovalbumin, aldolase, ferritin, chymotrypsinogen A	N-C_{18}	0.05 *M* P, pH 2, 10–50% 2-ME, 30 min, 2 ml/min	*38*
24. Insulin, β-lipotropin, BSA, neurotoxin 3, cytochrome *c*, ribonuclease, lysozyme, myoglobin	H-C_{18}	0.2 *M* P, pH 2.1, 0–60% CH_3CN, 1 h, 1 ml/min	*43*
25. Insulin, cytochrome *c*, and β-endorphin	μ-CN	0.2 *M* TEAP, pH 3, 12–33% CH_3CN, 30 min	*49*
26. Ovalbumin, prolactin, growth hormone, ribonuclease, calcitonin, carbonic anhydrase, elastase, β-lactoglobulin	U-SAC	0.155 *M* NaCl, pH 2.1, 0–75% CH_3CN, 90 min, 1 ml/min, 45°C	*39*
Other Proteins			
27. Neurophysins	RP18	0.01 *M* NaAc, pH 7, 51.6% CH_3OH, 1 ml/min	*11*
28. Neurophysins	μ-C_{18}	0.01 *M* NaP, pH 6, 40–50% CH_3OH, 90 min, 2.5 ml/h	*48*
29. Calcitonin-like proteins, hypothalamic extract	H-ODS	0.155 *M* NaCl, pH 2.1, 0–60% CH_3CN, 1 h, 1 ml/min	*40*
30. Calcitonins	μ-C_{18}	5 m*M* TBAP, pH 7.5, 20–75% CH_3OH, 20 min, 1.5 ml/min	*33*
31. α,β-Hemoglobin chains	S-C_{18}	0.36 *M* Pyr-formate, pH 3, 26% PrOH, 0.6 ml/min	*44*
32. α,β-Globin A chains, α,$\beta\gamma$,γ-globin F chains	μ-C_{18}	0.05 *M* P, pH 2.86, 42.5% CH_3CN, 1.5 ml/min	*58*
33. α,β,γ-Globin A chains	μ-C_{18}	0.15 *M* $NaClO_4$, 0.1% H_3PO_4, 5% CH_3OH, 15–75% CH_3CN, 80 min, 1.5 ml/min	*59*

34. HbA_1A_2 hemolysate	μ-C_{18}	0.1% P, 10–50% CH_3CN, 30 min, 2 ml/min	*24*
35. Interferon	L-C8	1 *M* NaAc, pH 7.5, 0–40% *n*PrOH, 3 h, 0.25 ml/min	*52*
36. Interferon	L-C8	1 *M* formic acid, 0.8 *M* Pyr, pH 4.2, 32–60% *n*PrOH, 40 min, 0.36 ml/min	*61*
37. Aldehyde dehydrogenase	μ-C_{18}	2 m*M* DDAP, pH 2.6, 25% *i*PrOH, 1.5 ml/min	*17*
38. Somatomedin C	μ-AP	0.01 *M* P, 25–40% CH_3CN, 1 h	*62*
39. Secretin	N-C_{18}	0.1% TFA, 15% CH_3OH, 1 ml/min	*68*
40. ACP	μ-AP	0.1% P, pH 2.1, 5% CH_3CN, 1.5 ml/min or 5 m*M* C_6SO_3Na, pH 6.5, 1.5 ml/min	*17*
41. Thyroid membrane glycoprotein	μ-C_{18}	0.1% P, pH 2.1, 10–50% CH_3CN, 30 min, 2 ml/min	*23*
42. Apolipoproteins C-I, C-II, C-III	μ-AP	1% TEAP, pH 3.2, 0–42% CH_3CN, 30 min, 1.5 ml/min	*15*
43. Apolipoproteins A-I, A-II	μ-AP or Zorbax C-8	1% TEAP, pH 3.2, 0–80% CH_3CN, 30 min, 1.5 ml/min	*20*
44. Apolipoproteins, C-I, C-II, C-III	μ-C_{18}	10 m*M* P, pH 6, 28–40% CH_3OH, 2 h, 1 ml/min	*54*
45. Ca^{2+}-binding proteins	μ-AP	10 m*M* P, pH 6.1, 5–65% CH_3CN, 20 min, 1.5 ml/min	*32*
46. Cytochromes *c*	μ-CN	0.1 *M* P, pH 2.0, 0.05 *M* Na_2SO_4, 27.5% CH_3CN, 2 ml/min	*65*
47. Aprotinine	N-C_{18}	0.1 *M* Na_2SO_4, pH 2.2, $CH_3OH : CH_3CN : H_2O$, 5 : 1 : 4, 1 ml/min	*9*
48. Collagen I, II, III	μ-CN	1.5 *M* Pyr-AcOH, pH 4.63, 0–10% *n*PrOH, 80 min, 0.3 ml/min	*8*
49. Parathyrin	μ-C_{18}	0.1% TFA or 0.13% HFBA, 0–80% CH_3CN, 80 min	*2*
50. β_2-Microglobulin	RP-C_{18}	0.012 *M* HCl, 35–80% EtOH, 30 min, 0.8 ml/min	*1*

[a] The following abbreviations have been used: L, LiChrosorb; μ-C_{18}, μ-Bondapak C_{18}; μ-AP, μ-Bondapak alkylphenyl; N, Nucleosil; μ-CN, μ-Bondapak alkylnitrile; H, Hypersil; S, Spherisorb; ES, ES Industries; RP, reversed phase; C_{18}, octadecyl-; C_8, octyl; C_6, cyclohexyl.

[b] The following abbreviations have been used: P, phosphate or phosphoric acid; TMAP, tetramethylammonium phosphate; TEAP, triethylammonium phosphate; TBAP, tetrabutylammonium phosphate; Pyr, pyridine; TFA, trifluoroacetic acid; HFBA, heptafluorobutyric acid; DDAP, dodecylammonium phosphate; 2-ME, 2 methoxylethanol.

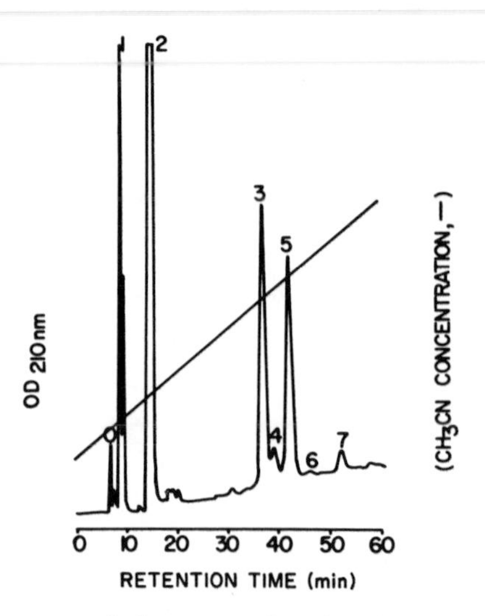

FIG. 15. The gradient reversed-phase separation of bovine and porcine insulin with 0.05 *M* tetramethylammonium phosphate, pH 3, as the mobile phase, and acetonitrile as the organic modifier. The column used was a LiChrosorb RP18 and the flow rate was 1 ml/min. The following peaks were identified: 1, hydroxybenzoic acid; 2, *p*-hydroxymethyl benzoate; 3, bovine insulin; 4, monodesamido bovine insulin; 5, porcine insulin; 6, monodesamido porcine insulin; 7, proinsulin. Adapted from Fig. 5 of Biemond *et al.* (1979).

tions a 0.1–0.2 *M* solution of an amine phosphate gives better peak shapes and recoveries than a dilute solution of phosphoric acid. As was described earlier, there are several explanations for the excellent results obtained with the amine salt, such as a general electrolyte effect, deactivation of silanol groups, and ion-pairing effects.

E. Optimum pH for Mobile Phase

The separation conditions described in Table V show that a low pH (2.5–3.5) is preferable for almost all protein separations. This pH value is compatible with reversed-phase columns and ensures that the silanol groups present in the packing are not ionized. For most proteins it is best to avoid a pH value near the isoelectric point of the protein to minimize the possibility of precipitation; e.g., insulin is insoluble at pH values near its p*I* of 5.5. Since all carboxylate anions present in proteins will be protonated at pH values of less than 3, the use of low pH values in reversed-phase HPLC should ensure that the protein samples are cationic and thus minimize any aggregation effects. Some proteins, however, may be unstable at low pH values and therefore higher pH values may be necessary. O'Hare and Nice (*43*) found that in the chromatography of ACTH with a 0.1 *M* NaH_2PO_4 buffer, an increase in pH from 2.1 to 4.5 resulted in perceptible broadening of the eluted peaks, and at pH 8.5 the peak became very broad. Calcitonin, however, did not show any pH effects when chromatographed under the same conditions. Alvarez *et al.* (*1*) found that β-microglobulin could be eluted from RP18 columns only if the pH of the mobile phase is low (2.5 or less).

The pyridine–acetic or formic acid system developed by Stein and co-workers (*61*) (see example 10 in Table V) has given results comparable to the amine phosphates. In this system fluorescent detection is required, but the method has the advantage that pyridine salts are volatile, which greatly simplifies isolation of the separated proteins. Again a high concentration of the salt (2 *M*) is used to deactivate silanol groups present in the column packing. This method uses somewhat higher pH values (3.5–5) than does the amine phosphate system and thus could be useful for the separation of proteins that are unstable at low pH values. Most protein samples will require careful handling when exposed to aqueous organic solvent mixtures at low pH values. For example, Biemond *et al.* (*3*) found that insulin was converted to the desamido derivative at a moderate rate (3%/day at pH 3). This side reaction could not be detected if fresh solutions were kept at 4°C and used on the same day. In the separation of apolipoproteins we found that eluted samples should be neutralized and freeze-dried immediately to avoid decomposition of the protein sample.

F. Summary of Use of Ion-Pairing Reagents

Clearly, the ability to dramatically increase or decrease the retention time of a protein on a reversed-phase column will prove to be invaluable in the analysis of complex biological mixtures. As well as changes in retention time, one can also obtain important changes in the selectivity of the column and thus in the observed elution order of the samples. The use of ionic and nonionic detergents such as SDS in the mobile phase could prove to be useful in the chromatography of membrane proteins and other materials which have a tendency to aggregate. Reversed-phase columns may also prove useful in the removal of detergents from protein samples, in view of the high affinity of the C_{18} reversed phase for such modifiers.

V. EXAMPLES OF PROTEIN SEPARATIONS

Table V lists a selection of protein purifications which have been achieved by reversed-phase HPLC. Figure 5 shows that a wide range of

TABLE VI

Recoveries of Protein Samples after Separation by Reversed-Phase HPLC

Protein sample	Recovery (%)	Separation conditions	Protein sample	Recovery (%)	Separation conditions
Tyrosinase, collagen α_1, cytochrome *c* bovine serum albumin	>80[b]	22	Ribonuclease, lysozyme, myoglobin, globin chains	70–100	32
ACTH	>90[c]	13	Somatomedin C	>80	38
Calcitonin-like materials	>80	29[e]	Apolipoproteins C-I, C-II, C-III	97, 85.5, 83	42
Insulin, β-lipotropin, albumin, neurotoxin, cytochrome *c*	50–80[d]	24	Cytochromes *c*	>80	46
			Growth hormone	61	21
			Parathyroid hormone	79	18

[a] Chromatographic conditions as in Table V; number refers to examples in this table.
[b] For injection of 10-μg samples.
[c] Measured by radioimmunoassay and corticosteroid release from a suspension of isolated cells.
[d] Lower recoveries observed for small sample loading (1–5 μg).
[e] Used 0.155 *M* NaCl, pH 2.1, as mobile phase because phosphate salts not compatible with the bioassay.

polypeptides can be successfully purified by the technique, although to date few enzymes have been chromatographed with quantitation of the recovered enzymic activity. Table VI shows that, provided suitable separation conditions can be developed, excellent recoveries of the purified protein can be obtained. Although octadecyl (C_{18}) is a popular choice for the reversed-phase column, many separations are carried out on columns of lower hydrophobicity, such as octyl or phenyl packings. Mobile phases which contain phosphate salts of an amine are widely used in combination with acetonitrile or a short-chain alcohol as organic modifier. An important advantage of this mobile-phase system is that it is compatible with low-wavelength UV detection at 215 nm. This wavelength is particularly suitable for detection of proteins, because it corresponds to an isobestic point at which random and helical peptide bond absorptions are equal (*12*). At this wavelength the UV absorption is independent of conformation and, ignoring absorption tails of aromatic residues, the molar absorptivity is 10^3 $M^{-1}cm^{-1}$ per residue. A useful alternative is the pyridine–acetic or formic acid system (see Table V) in combination with fluorescent rather than UV detection of the effluent. Although Table V can be used as a guide to general separation conditions, it is likely that a new protein separation will require a unique set of chromatographic conditions.

Acknowledgments

This material was developed by the National Research and Demonstration Center, National Heart, Lung and Blood Institute, National Institutes of Health, Grant. No. HL-17269. Support is also acknowledged from The National Heart Foundation and Medical Research Council of New Zealand (WSH). JTS is an Established Investigator of The American Heart Association.

REFERENCES

1. V. L. Alvarez, C. A. Roitsch, and O. Henriksen, *Anal. Biochem.* **115,** 353–358 (1981).
2. H. P. J. Bennett, S. Solomon, and D. Goltzman, *Biochem. J.* **197,** 391–400 (1981).
3. M. E. F. Biemond, W. A. Sipman, and J. Olivie, *J. Liq. Chromatogr.* **2,** 1407–1435 (1979).
4. C. A. Browne, H. P. J. Bennett, and S. Solomon, *Biochem. Biophys. Res. Commun.* **100,** 336–343 (1981).
5. C. A. Browne, H. P. J. Bennett, and S. Solomon, *Biochemistry* **20,** 4538–4546 (1981).
6. J. S. D. Chan, N. G. Seidah, C. Gianoulakis, A. Bélanger, and M. Chrétien, *J. Clin. Endocrinol. Metab.* **51,** 364–367 (1980).
7. A. Dinner and L. Lorenz, *Anal. Chem.* **51,** 1872–1873 (1979).
8. A. Fallon, R. A. Lewis, and K. D. Gibson, *Anal. Biochem.* **110,** 318–322 (1981).
9. M. Gazdag and G. Szepesi, *J. Chromatogr.* **218,** 603–612 (1981).
10. S. Gentleman, L. I. Lowney, B. M. Cox, and A. Goldstein, *J. Chromatogr.* **153,** 274–278 (1978).
11. J. A. Glasel, *J. Chromatogr.* **145,** 469–472 (1978).
12. W. B. Gratzer, *in* "Poly-α-amino acids, Biological Macromolecules Series" (G. D. Fasman, ed.), pp. 177–238. Dekker, New York, 1967.
13. I. Halasz, *Anal. Chem.* **53,** 1393–1396 (1980).
14. W. S. Hancock, C. A. Bishop, J. E. Battersby, D. R. K. Harding, and M. T. W. Hearn, *J. Chromatogr.* **168,** 377–384 (1978).
15. W. S. Hancock, C. A. Bishop, A. M. Gotto, D. R. K. Harding, S. M. Lamplugh, and J. T. Sparrow, *J. Lipid Res.* **16,** 250–259 (1981).
16. W. S. Hancock, C. A. Bishop, L. J. Meyer, and D. R. K. Harding, *J. Chromatogr.* **161,** 291–298 (1978).
17. W. S. Hancock, C. A. Bishop, R. L. Prestidge, D. R. K. Harding, and M. T. W. Hearn, *Science* **200,** 1168–1170 (1978).
18. W. S. Hancock, J. D. Capra, W. A. Bradley, and J. T. Sparrow, *J. Chromatogr.* **206,** 59–70 (1981).
19. W. S. Hancock and J. T. Sparrow, *in* "A Laboratory Manual for the Separation of Biological Materials by HPLC." Dekker, New York (in preparation).
20. W. S. Hancock, H. J. Pownall, A. M. Gotto and J. T. Sparrow, *J. Chromatogr.* **216,** 285, 1981.
21. R. H. Haschemeyer and A. E. Haschemeyer, *in* "Proteins: A Guide to Physical and Chemical Methods," p. 352. Wiley, New York, 1973.
22. M. T. W. Hearn, *Adv. Chromatogr.* **18,** 59–81 (1980).
23. M. T. W. Hearn and W. S. Hancock, *J. Liq. Chromatogr.* **2,** 217–239 (1979).
24. M. T. W. Hearn and W. S. Hancock, *Chromatogr. Sci.* **10,** 243–272 (1979).
25. M. T. W. Hearn and W. S. Hancock, *Trends Biochem. Sci.* **4,** N58–N62 (1979).
26. S. Hjertén, *J. Chromatogr.* **87,** 325 (1973).

27. B. H. J. Hofstee, *in* "Methods of Protein Separation" (N. Catsimpoolas, ed.), Vol. 2, Chapter 6, p. 233. Plenum, New York, 1976.
28. C. Horváth and W. Melander, *J. Chromatogr. Sci.* **15,** 393–404 (1977).
29. C. Horváth, W. Melander, and I. Molnar, *J. Chromatogr.* **125,** 239–256 (1977).
30. N. Kangawa, N. Minamino, N. Chino, S. Sakakibara, and H. Matsuo, *Biochem. Biophys. Res. Commun.* **99,** 871–888 (1981).
31. W. Kissing and R. H. Reiner, *Chromatographia* **11,** 83 (1978).
32. C. B. Klee, M. D. Oldewurtel, J. F. Williams, and J. W. Lee, *Biochem. Int.* **2,** 485–493 (1981).
33. P. W. Lambert and B. A. Roos, *J. Chromatogr.* **198,** 293–299 (1980).
34. R. V. Lewis, A. Fallon, S. Stein, K. D. Gibson, and S. Udenfriend, *Anal. Biochem.* **104,** 153–159 (1980).
35. J. L. Meek, *Proc. Natl. Acad. Sci. U.S.A.* **77,** 1632–1636 (1980).
36. J. Meienhofer, T. F. Gabriel, J. Michalewsky, and C. H. Li, "Peptides." Wroclaw Univ. Press, Poland, 1979.
37. I. Molnar and C. Horváth, *J. Chromatogr.* **142,** 623–640 (1977).
38. W. Monch and W. Dehnen, *J. Chromatogr.* **147,** 415–418 (1978).
39. E. C. Nice, M. W. Capp, N. Cooke, and M. J. O'Hare, *J. Chromatogr.* **218,** 569–580 (1981).
40. E. C. Nice, M. Capp, and M. J. O'Hare, *J. Chromatogr.* **147,** 413–427 (1979).
41. E. C. Nice and M. J. O'Hare, *J. Chromatogr.* **162,** 401–407 (1979).
42. M. J. O'Hare and E. C. Nice, *J. Chromatogr.* **149,** 241 (1978).
43. M. J. O'Hare and E. C. Nice, *J. Chromatogr.* **171,** 209–226 (1979).
44. P. E. Petrides, R. T. Jones, and P. Bohlen, *Anal. Biochem.* **105,** 383–388 (1980).
45. J. Poráth, L. Sundberg, N. Fornstedt, and I. Olsson, *Nature* (*London*) **245,** 465 (1973).
46. H. J. Pownall, J. B. Massey, and A. M. Gotto, Jr., in press.
47. R. Rekker, "The Hydrophobic Fragment Constant," p. 301. Elsevier, Amsterdam, 1977.
48. W. Richter and P. Schwandt, *J. Neurochem.* **36,** 1279–1280 (1981).
49. J. E. Rivier, *J. Liq. Chromatogr.* **1,** 343–366 (1978).
50. M. Rubinstein, S. Rubinstein, P. C. Familletti, R. S. Miller, A. A. Waldman, and S. Pestka, *Proc. Natl. Acad. Sci. U.S.A.* **76,** 640–644 (1979).
51. M. Rubinstein, S. Stein, L. D. Gerber, and S. Udenfriend, *Proc. Natl. Acad. Sci. U.S.A.* **74,** 3052–3055 (1977).
52. M. Rubinstein, S. Stein, and S. Udenfriend, *Proc. Natl. Acad. Sci. U.S.A.* **74,** 4969–4972 (1977).
53. G. Schill, *Ion Exch. Solvent Extr.* **6,** 1 (1974).
54. P. Schwandt, W. Richter, and P. Weisweiler, *J. Chromatogr.* **225,** 185–188 (1981).
55. R. P. W. Scott and P. Kucera, *J. Chromatogr.* **142,** 213–232 (1977).
56. J. P. Segrest, R. L. Jackson, J. D. Morrisett, and A. M. Gotto, *FEBS Lett.* **38,** 247–252 (1974).
57. N. G. Seidah, R. Rothier, S. Benjannet, N. Lariviere, F. Gossard, and M. Chrétien, *J. Chromatogr.* **193,** 291–299 (1980).
58. J. B. Shelton, J. R. Shelton, and W. A. Schroeder, *Hemoglobin* **3,** 353–358 (1979).
59. J. B. Shelton, J. R. Shelton, and W. A. Schroeder, *J. Liq. Chromatogr.* **4,** 1381–1392 (1981).
60. S. J. Singer and B. Nicholson, *Science* **175,** 720 (1972).
61. S. Stein, C. Kenny, H.-J. Friesen, J. Shively, U. Del Valle, and S. Pestka, *Proc. Natl. Acad. Sci. U.S.A.* **77,** 5716–5719 (1980).

62. M. E. Svoboda, J. J. van Wyk, D. G. Klapper, R. E. Fellows, F. E. Grisson, and R. J. Schleuter, *Biochemistry* **19,** 790–797 (1980).
63. G. Szepesi and M. Gazdag, *J. Chromatogr.* **218,** 597–602 (1981).
64. S. Terabe, R. Konaka, and K. Inouye, *J. Chromatogr.* **172,** 163–177 (1979).
65. S. Terabe, H. Nishi, and T. Ando, *J. Chromatogr.* **212,** 293 (1981).
66. K. Unger and R. Kern, *J. Chromatogr.* **122,** 345 (1976).
67. J. W. Van Nispen, W. A. A. J. Bijl, and H. M. Greven, *Rec. Trav. Chim. Pays-Bas* **99,** 57–62 (1980).
68. D. Voskamp, C. Olieman, and H. C. Beyerman, *Rec. Trav. Chim. Pays-Bas* **99,** 105–108 (1980).
69. J. M. Zanelli, M. J. O'Hare, E. C. Nice, and P. H. Corran, *J. Chromatogr.* **223,** 59–67 (1981).

HIGH-PERFORMANCE LIQUID CHROMATOGRAPHY OF PEPTIDES

Milton T. W. Hearn

St. Vincent's School of Medical Research
Melbourne, Victoria, Australia

HIGH-PERFORMANCE LIQUID
CHROMATOGRAPHY, Vol. 3

ISBN 0-12-312203-1

I. INTRODUCTION

There can now be little doubt that during the last decade, peptide chemistry finally came of age after more than 50 years of development. Its long and tenuous childhood may be considered to have commenced with Sumner's demonstration (*1*) in 1926 that the enzyme urease was a protein. It showed the first signs of maturity with the innovative synthetic studies on oxytocin and its analogs by du Vigneand *et al.* (*2*) and the definitive elucidation of the primary structure of bovine insulin by Sanger *et al.* (*3*). Peptide chemistry now occupies a pivotal place germane to most areas of the life sciences. Much of our current understanding of the molecular pathways operating in endocrinology and immunology, to name just two areas, has been dependent upon techniques developed in peptide chemistry. Many of the biologically active peptides, polypeptides, and proteins now being studied, e.g., regulation and growth factors, membrane-associated proteins, multicomponent enzymes, and neuroendocrine hormones, are present in animal or plant tissues at concentrations below a microgram per gram of tissue. For their analysis and purification, the peptide chemist requires a number of sophisticated methods. Experience has taught him that to be of practical value, these methods must be selective, ideally specific, and sensitive.

Biological *in vivo* or *in vitro* response assays, as now routinely used, can detect a particular peptide or protein at a level of sensitivity several orders of magnitude greater than can be obtained with classical procedures for chemical analysis. However, biological assays can be demanding experimentally and are often expensive. They have been supplanted where possible by radioligand or radioimmunoassays which permit detection in favorable circumstances well below the picomole level. Such functional assays generally provide little information on the precise molecular characteristics of the peptide or protein. This uncertainty can lead to poorly justified conclusions on the molecular relatedness of different molecules. An excellent example of the need for cautious interpretation of binding data based on immunological or biological criteria has been the recent demonstration that the high-molecular-weight immunoreactive "β-endorphin" present in extracts of human placenta is related to the heavy chain of immunoglobulin IgG_1 (*4*). In practically all cases, functional assay procedures are dependent on the availability of highly purified, and well-characterized, standard substances to allow relative potencies to be assessed, specific antibodies to be raised, or binding characteristics to be compared. The detailed chemical characterization of natural, as well as synthetic, peptide substances thus remains a necessary prelude to any discussion on their fundamental biological function and significance. Whether the study involves the isolation, the purification and characterization, the sequence elucidation, total or semisynthetic forays, or the determination of structure–function relationships, selective methods of separation capable of exploiting the compositional and structural features of the peptidic molecule under study are required.

Electrophoretic techniques, such as two-dimensional electrophoresis or isoelectric focusing, lend themselves under appropriate conditions to the separation with excellent resolution of small amounts of samples. They have mainly found use as powerful methods of analysis since their general applicability in the areas of peptide purification and isolation has, until recently, been severely restricted by limitations in sample capacity and instrumental design.

The need for rapid and versatile liquid chromatographic techniques which would provide a level of resolving power similar to that obtained by the best electrophoretic procedures, has long been recognized in peptide chemistry. During the last decade, column liquid chromatography has made a number of important advances in this direction with the introduction of affinity chromatography, partition chromatography, hydrophobic interaction chromatography, and, most recently, high-performance liquid chromatography (HPLC). It is now clearly apparent that HPLC techniques, whether used alone or in conjunction with alternative open column procedures, will increasingly influence the direction of most areas of study involving peptides and related biomolecules. These favorable circumstances have come about due to two significant developments. The first advance has come from the recognition that column efficiency and speed of separation could be enhanced by several orders of magnitude if pressure-stable support materials of defined surface chemical properties and small particle size (e.g., microparticulate porous silicas of nominal particle diameter d_p equal to 5 or 10 μm with accessible surface silanols chemically bonded to alkyl chains) were used as column packings. The second development relates to the use of a greatly extended variety of elution conditions, many of which are novel to the field of peptide separation, for the optimization of resolution with a particular type of packing material. Because a number of fundamentally important parameters—nature of the packing material, composition and pH of the mobile phase, flow rate and temperature—can be readily varied with the same basic HPLC equipment, a high resolving power and a high speed of separation for solutes can be simultaneously obtained with these chromatographic systems.

Modern HPLC separations of peptides now encompass all of the chromatographic modes—liquid–solid adsorption, liquid–liquid partition, ion-exchange and gel permeation chromatography—with the reversed-phase HPLC approaches currently attracting the greatest attention. The versatility of these methods has been demonstrated in a wide variety of studies with natural and synthetic peptides covering such divergent areas as the isolation of putative enkephalin precursors from bovine adrenal mellula (*5*), the tryptic mapping of thyroid antigenic proteins (*6*), the resolution of diastereomeric peptides (*7*), the assessment of homogeneity and purification of synthetic protected and deprotected peptides of pharmaceutical importance (*8, 9*). The considerable impact which modern liquid chromatographic techniques is having on all areas of peptide and protein chemistry, and in particular the biomedical disciplines, was highlighted at recent International Symposia in Washington, D.C. and Baltimore and extensively documented in the associated Proceedings Volumes (*9a,b*).

This article will deal chiefly with the separation of peptides and polypeptides by reversed-phase HPLC. The choice of peptides and polypeptides has been deliberately limited by the somewhat arbitrary distinction of a molecular weight cutoff of ca. 20,000. The separation of amino acids and polypeptides—proteins with MW > 20,000—has recently been reviewed elsewhere (*10–13a,b*). Although peptides have traditionally been separated on ion-exchange or gel permeation chromatographic systems, recent experiences with peptide separations have confirmed that high resolution, good recoveries and considerable experimental flexibility can now be anticipated for the stable reversed phases supported on inert microparticles. Before we consider the various applications of modern HPLC with these reversed phases, as well as with the other types of chromatographic phases now available and the impact these methods are having on peptide chemistry, it is essential to briefly examine the theoretical framework upon which these HPLC techniques are built.

II. GENERAL CHROMATOGRAPHIC CONSIDERATIONS

The aim of any chromatographic separation may be defined as the achievement of an optimal combination of speed of elution, sample size, and resolution of the solutes. Good resolution can only be obtained if there is adequate control over the differential migration rates of a group of solutes as they move down a column (column selectivity) and over the extent of zone dispersion for each of the solutes (column efficiency). Historically, the various modes of liquid chromatography have been considered as separate and independent phenomena. It is now clear that they all have a common theoretical basis. Column selectivity in HPLC, irrespective of the mode, arises due to differences in the distribution equilibria established by the various solutes between the stationary and mobile phases whereas zone dispersion originates in multiple flow paths, in axial molecular diffusion, and in the kinetics of desorption from the stationary phase. Consequently, solute retention, which is usually expressed in terms of the capacity factor k', is governed by thermodynamic considerations whereas zone dispersion or band spreading, which is most conveniently expressed in terms of the reduced plate height h, is essentially a kinetic phenomenon. It can be readily shown that resolution R is related to efficiency, selectivity α, and the capacity factor k' by

$$R = \frac{1}{4}\left(\frac{L}{hd_p}\right)^{1/2} \frac{\alpha - 1}{\alpha} \frac{k'}{k' + 1} \tag{1}$$

where L is the column length, h the reduced plate height, and d_p the nominal particle diameter.

In common with other application areas of chromatographic separation, a considerable amount of effort has been expended recently on the development of different elution conditions and types of stationary phases for peptide separations in attempts to maximize column selectivities without adversely affecting column efficiences. Peptide retention will invariably be mediated by the participation of electrostatic, hydrogen bonding, and hydrophobic interactions in the distribution phenomenon. The nature of the predominant distribution mechanism will be dependent on the physical and chemical characteristics of the stationary phase as well as the nature of the molecular forces which hold the solute molecules within the mobile and stationary zones. The retention of the solute in all HPLC modes can be described by the equation

$$V'_{R,i} = K_i A'_{s,i} \tag{2}$$

where $V'_{R,i}$ is the corrected retention volume of the solute, K_i is the distribution coefficient of the solute, and $A'_{s,i}$ can be the surface area or the volume of the stationary phase depending on the dominant chromatographic mode. High affinity of the peptidic solute for the stationary phase will be manifested as large values of K_i whereas for a true nonionic steric exclusion mode, K_i values will range between 0 and 1.

In steric exclusion (gel permeation) separations, the retention of partially retained solutes is described by

$$V_{R,i} = V_0 + K_i V_I \tag{3}$$

where V_0 is the void volume and V_1 the total pore volume. In view of the linear dependency of $V_{R,i}$ on the logarithm of molecular weight over a fairly wide range of molecular weights, selectivity will be controlled by the slope n of the plot V_R versus log MW such that

$$n = \frac{\Delta \log \mathrm{MW}}{\Delta V_{\mathrm{R},i,j}} \tag{4}$$

where Δlog MW is the molecular weight difference between two solutes S_i and S_j, and $\Delta V_{\mathrm{R},i,j}$ is the elution volume difference. As will be discussed subsequently (see page 140), the current range of microparticulate silicas suitable for gel permeation HPLC, do not permit sufficient selectivity differences to be of general use for the separation of polypeptides with MW < 5000.

For ion-exchange HPLC, the retention of a cationic peptidic solute of charge z can be related to the ionic strength of the eluent and the pH such that

$$k_i' = \frac{\Phi c}{[\mathrm{El_{aq}}]^z + ([\mathrm{El_{aq}}]^z[\mathrm{H_{aq}}^+]/K_{\mathrm{av},i})} \tag{5}$$

where Φ, $[\mathrm{El_{aq}}]$, $K_{\mathrm{av},i}$ are the phase volume, the ionic strength in the eluent, and the average distribution coefficient for the solute in its various mono-, di-, . . . , polycationic forms, respectively, and c is a system constant. Major selectivity differences are expected, on the basis of Eq. (5), to arise from small changes in pH, $[\mathrm{El_{aq}}]$, or Φ. The use of ion-exchange HPLC for polypeptide separation is discussed on page 143.

With polar liquid–liquid adsorption chromatography, based on chemically bonded normal-phase systems, the distribution coefficient can be equated with the solubility parameter δ_i of a solute such that retention is given by

$$\log k_i' = \log \Phi + \frac{V_i[(\delta_i - \delta_\mathrm{m})^2 - (\delta_i - \delta_\mathrm{s})^2]}{RT} \tag{6}$$

where δ_m and δ_s are the solubility parameters of the mobile and stationary phases, respectively, and V_i is the molar volume of solute i. Thus, as the affinity of a peptide for the polar stationary phase increases, δ_i becomes more like δ_s and a more polar eluent is required to affect elution. The reverse applies as the affinity of the peptide for the stationary phase decreases and the polarity of solute i becomes lower. The major use of polar liquid–liquid adsorption HPLC is for the normal-phase separation of peptides of low polarity, i.e., very hydrophobic underivatized peptides, fully protected synthetic peptides, etc. Both the chemically bonded normal- and reversed-phase silicas, under appropriate conditions, can be coated by solvent molecules in the mobile phase. This coating process will generate partition–adsorption phase systems formally analogous to the more familiar liquid–liquid partition systems previously used in the open column separation of peptides.

The equilibrium process describing the adsorption and desorption of a solute i in a liquid–solid chromatographic separation can be expressed by

$$\log K_{ads,i} = \log V_a + \beta(S^0 - A_i\epsilon^0) \tag{7}$$

where $K_{ads,i}$ is the adsorption coefficient of solute i and is the ratio of the moles of solute adsorbed per gram of sorbent to the moles of solute per cubic centimeter of eluent; V_a is the volume of an adsorbed monolayer of solvent per gram of sorbent with a monolayer thickness of ca. 0.35 nm; β is the surface activity parameter of the sorbent with an arbitrary value of unity for a thermally activated, i.e., "dry," surface; $S°$ is the dimensionless free energy of adsorption of the solute onto the surface; A_i is the molecular area of solute i in units of 0.085 nm^2; and $\epsilon°$ is the solvent strength parameter relative to pentane, i.e., $\epsilon°_{pentane} = 0$. With amphoteric solutes, poor reproducibility and resolution is often experienced with silica supports due to strong solute–sorbent interactions and the variability of the V_a and β terms when polar mobile phases are used. With a given mobile phase, selectivity will only be slightly influenced by changes in the surface area of the sorbent but will show dramatic variations due to small changes in the concentration of the water or other protic components in the mobile phase. Because of these effects, which are often difficult to control experimentally with polar solutes, this chromatographic mode has almost exclusively been used only for the separation of fully protected peptides.

Solute retention in reversed-phase HPLC is dependent on the different distribution coefficients established between a polar mobile and a nonpolar stationary phase by the peptidic components of a mixture. Although there are many similarities between reversed-phase HPLC separations of peptides and the classical liquid–liquid partition chromatographic methods, it is debatable whether the sorption process in reversed-phase HPLC arises due to partition or adsorption events, i.e., whether the nonpolar stationary phase functions as a bulk liquid or as an adsorptive monolayer. These aspects and the theoretical models for reversed-phase HPLC are discussed in a subsequent section.

We now have a fairly adequate understanding of the different properties, including the particle diameter d_p, the pore size, the degree of permeability, and the chemical composition of the surface of the support matrix, to know which type of stationary phase can be successfully used with a particular class of peptides. Most of the HPLC packing materials now in use for peptide separations are based on the wide pore microparticulate silica gels with polar or nonpolar carbonaceous phases chemically bonded to the surface of the matrix. Methods for the preparation of these chemically bonded stationary phases, their available sources of supply, and the practical aspects of column packing and handling have been the subjects of a number of recent reviews and monographs (*14–16*). The newer, fully porous chemically modified packing materials, derived from silica aerogels and other types of fused aggregates, exhibit column efficiencies at least an order of magnitude higher than the earlier pellicular supports at comparable sample loadings and speed of separation. Most recent applications of HPLC separations of peptides have concentrated on the use of irregular or spherical porous silica particles with mean particle diameters in the range 5–10 μm, surface areas of ca. 30–800 m^2/g, and nominal pore diameters of 5–50 nm. Silica matrices with larger pore sizes have attracted particular interest in gel permeation and ion-exchange HPLC, where solute permeability is an important consideration. Very high separation efficiencies can be obtained with the microparticulate supports for which d_p = 5–10 μm. As a consequence, a standard stainless steel analytical column (25 × 0.4 cm) packed with, for example, a uniform alkyl silica with d_p = 10 μm will generally exhibit sample capacities encompassing the analytical (1 ng to 100 μg) and semipreparative (10 μg to 10 mg) separation range. Larger-particle-diameter alkyl silica supports (e.g., d_p = 75 μm) have found application for preparative (>1 gm) purifications of synthetic peptides and in batch isolations of biological peptides. The major limitation of silica-based supports is their chemical instability at high pHs. Below pH 7.5, they show good stability over a wide range of elution conditions, and because of their particle size and mechanical strength, they can be used over a wide range of linear flow velocities.

Not unexpectedly, considerably less information is available at this stage on the precise nature and magnitude of the molecular forces involved in the interactions between peptides and the mobile and the stationary phases currently used in HPLC separations. As a consequence, it is not yet possible to predict all of the subtle selectivity differences that may arise when one mobile phase condition is substituted for another. Illustrative of this point is the elution order reversal observed (*17*) with the separation on LiChrosorb ODS of insulin and $ACTH_{1-39}$ when the organic solvent modifier (used in this reversed-phase separation) was changed from methanol to acetonitrile. However, recent systematic studies (*17a–17c*) on the influence of the stationary-phase surface area and porosity and on the effect of phase volume changes of different organic solvent modifiers have provided important new insights into the chromatographic phenomena involved with peptide selectivity on chemically modified microparticulate silicas. Despite the current interpretative limitations, the recognition that peptide retention in the various HPLC separative modes can be accommodated by functional group additivity approaches has proved most successful in establishing criteria for the selection and optimization of elution conditions. This treatment has, in particular, proved of considerable value in evaluating selectivity effects in the reversed-phase HPLC separation of unprotected peptides.

III. SEPARATION OF PEPTIDES ON CHEMICALLY BONDED REVERSED PHASES

A. Historical Background

The maintenance of tertiary and quarternary structure of polypeptides and proteins, as well as their mechanisms of association with other cellular components, is known to be, in part, mediated by hydrophobic interactions. In fact, the most important single factor in the organization of the constituent molecules of living matter into complex structural entities and the subsequent transmission of the encoded information from one structure to another is probably the hydrophobic effect. Consequently, it is not surprising that separative methods should have been developed which exploit this phenomenon.

Hydrophobic interaction chromatography as introduced by Porath *et al.* (*18*) involves the separation at low pressures of proteins and other polyelectrolytes on amphiphilic gels prepared by immobilizing hydrocarbonaceous ligands such as alkyl or aryl groups onto soft hydrophilic gels. With these amphiphilic gels the immobilization procedure does not introduce ionic groups in contrast to similar adsorbents prepared by coupling long-chain aliphatic amines to cyanogen bromide-activated agarose (*19, 20*). The recently introduced CDI-activation approach (*21–21b*) for biospecific, ligand-specific, or hydrophobic interaction affinity chromatography also does not introduce additional ionic groups into a matrix. With these and the Porath gels, solute retention is mainly due to nonpolar interactions between the solute molecules and the nonionic amphiphilic stationary phase instead of the mixed ionic–hydrophobic adsorption of the CNBr-activated agarose derivatives. These gel types are primarily intended for open-column separation of polypeptides and proteins via the sorptive interactions between the accessible hydrophobic domains on the surface of the polyelectrolyte and those present on the nonpolar ligand. Desorption can be achieved by decreasing the elutropic strength of the mobile phase, i.e., by lowering the ionic strength; by changing the buffer ions to those of greater chaotropicity; by varying the pH; or by reducing the polarity of the eluent via the addition of alcohols or detergents.

Because agarose and derived amphiphilic supports show low efficiencies, they have attracted little interest for the high-resolution separation of peptides based on their relative hydrophobicities. Separation of this type can, however, be achieved by liquid–liquid partition chromatography. The partition concepts developed by Martin and Synge (*22*) were successfully applied by Yamashiro (*23*) to the open-column chromatographic purification of oxytocin on a Sephadex G-25 support. This method has subsequently been widely used for the separation of peptides on the basis of hydrophobicity differences and includes the insulin C peptides (*24*), analogs of somatostatin (*25*), and β-endorphins (*26*). These partition systems have usually been based on 1-butanol–acetic acid–water (BAW) or 1-butanol–acetic acid–water–pyridine (BAWP) combinations and the small/medium-pore, cross-linked dextrans such as Sephadex G-50 or LH-60. Selectivity differences, which reflect the relative hydrophobicities of the peptides, can be related to the incremental free energy changes arising from variations in the peptide chain length or the replacement of one amino acid side chain by another. Although useful peptide separations can be obtained with these classic partition chromatographic methods, they exhibit only modest column efficiencies, e.g., ca. 500–1000 theoretical plates per meter, and are time consuming. In addition, these methods are incompatible with gradient elution and this limitation restricts the polarity range of peptides that can be separated by any one phase system.

The advent of the chemically bonded reversed phases based upon silica microparticles with d_p = 5 or 10 μm has circumvented the disadvantages of the above two methods. Since a nonpolar moiety, usually an octyl or a octadecyl group, is chemically bonded to the silica surface, gradient elution techniques can be used and there is also much more flexibility in the choice of the composition of the mobile phase. With current bonding technology, it is not possible to react all the free silanol groups (about 10 μmol/m^2) on the surface of the silica matrix particle. Surface coverages of alkyl ligands of ca. 15 μmol/m^2 have been reported although generally the coverage is in the range 3–9 μmol/m^2 depending on the choice of primary and secondary silanization reagents used. Besides the intrinsic residual level of free silanol groups, an additional source arises when monochlorosilane reagents are contaminated with di- and trichlorosilanes, which leads to the formation of a weakly bound polymeric layer. The progressive regeneration of silanols due to leaching of the polymeric layer will lead to column instability, poor reproducibility, and loss of resolution. For these reasons, alkyl silicas prepared with high-purity monochlorosilanes, end-capped under forcing reaction conditions, and subsequently subjected to an extensive washing and equilibration procedure are recommended for reversed-phase HPLC of peptides.

Following the initial observations reported (*27, 27a*) in 1976 on the separation of a series of unprotected peptides, related to the hypothalamic-releasing factors and angiotensin family, using pellicular and fully porous microparticulate alkyl silicas, and aquo-organic solvent–eluent combinations at low pH, it was anticipated that reversed-phase HPLC approaches would allow the rapid and selective separation of unprotected and protected peptides at both the analytical and preparative levels. Subsequent studies have confirmed this anticipated versatility.

B. Ligand Adsorption, Partition, and Solvophobic Considerations

One of the most favorable aspects of the use of reversed-phase HPLC for peptide separation is that neat or partially aqueous mobile phases are employed. Under elution conditions where peptides exhibit regular reversed-phase behavior, e.g., with water-rich eluents of low pH, low ionic strength, and in the absence of hydrophobic pairing ion interaction or micelle formation, the most polar molecule will have the weakest interaction with the hydrophobic stationary phase and thus the smallest elution volume. The involvement of secondary solution equilibria, notably changes in ionization state and buffer pairing ion interactions, will significantly modulate the retention behavior of a particular peptide. With small peptides, the retention process can be envisaged as involving interactions between a single class of equivalent domains on the surface of the peptide and the hydrocarbonaceous ligand. Larger polypeptides and proteins probably intercalate with the bonded ligand via multisite hydrophobic and polar surface interactions. When they occur, multisite interactions between the solute and the stationary phase will significantly influence the energetics and kinetics of the desorption process.

Several fundamental models have recently been proposed (*11, 17a,b, 28*) in order to quantitatively accommodate the influence of solution equilibria on the distribution process. Although application of these theoretical approaches has yet to provide a comprehensive general description of retention behavior for peptides, they have provided useful insight into those mobile-phase effects arising from changes in the nature and mole fraction of the organic solvent modifier.

The basic model for the separation of peptidic solutes on nonpolar stationary phases assumes that reversible interactions of the solute molecules $S_1, S_2, \ldots, S_n$ occur with the hydrocarbonaceous ligand L and that the interactions are due to hydrophobic associations and not to electrostatic or hydrogen bonding effects. Conceptually, the sorption of peptides to alkyl-bonded reversed phases under these conditions can be based either on partition or on adsorption processes. In a partition process, solute retention is related to the stationary-phase volume via the partition coefficient P, whereas in an adsorption process, solute retention is related to the total interfacial surface area via the adsorption coefficient K_A. For both processes, solute retention will be governed by the distribution equilibria established between the stationary and the mobile phases and characterized by the equilibrium constants $K_1, K_2, \ldots, K_n$, which in the general case can be defined as

$$K_i = [S_iL]/[S_i][L] \tag{8}$$

Solute retention is usually expressed in terms of the capacity factor k', and this can be directly related to the equilibrium constant using Eq. (2). The dependence of the capacity factor k'_i of a solute S_i on K_i is thus given by

$$k'_i = \Phi K_i \tag{9}$$

where Φ is the phase ratio of the system.

The thermodynamic equilibrium constants are related to the overall standard unitary free energy changes associated with the transfer of the solutes from the mobile to the stationary phase such that

$$\log K_i = \Delta G^{\circ}_{\text{assoc},i}/RT \tag{10}$$

and hence solute retention can be expressed as

$$\log k'_i = \log \Phi - \Delta G^{\circ}_{\text{assoc},i}/RT \tag{11}$$

and chromatographic selectivity $\alpha_{i,j}$ for two peptidic solutes S_i and S_j by

$$RT \log \alpha_{i,j} = \Delta(\Delta G^{\circ}_{\text{assoc},i,j}) \tag{12}$$

The log $\alpha_{i,j}$ term reflects the differences in capacity ratios of the two peptide solutes S_i and S_j which differ by a functional group and is analogous to the ΔR_m term used to predict selectivity differences for the classical liquid–liquid partition chromatography of peptides. The influence of functional group behavior on the retention of polar solutes in reversed-phase HPLC has been the subject of several recent articles and similar trends are apparent with peptide derivatives (*29–31*).

Because the physicochemical basis of chromatographic retention on hydrocarbonacous stationary phases with aqueous eluents and the liquid–liquid partition of solutes between nonpolar organic solvents, e.g., n-octanol, and aqueous solvents are similar, selectivities can be expressed in terms of partition coefficients by the relationship

$$\alpha_{i,j} = P_i/P_j \tag{13}$$

where P_i and P_j are the partition coefficients of the solutes S_i and S_j, respectively, for that system.

Partition coefficients for liquid–liquid distribution processes are generally related to the standard reference system, octanol–water, such that

$$\log P_{m,i} = z_m \log P_{\text{octanol-water},i} + C \tag{14}$$

where $P_{m,i}$ is the partition coefficient of the solute S_i in the mobile-phase system m, z_m is the eluotropic parameter, and $P_{\text{octanol-water},i}$ $(P_{o,i})$ is the partition coefficient of S_i in an octanol–water system.

Rekker (*32*) has derived, from a modified version of the Hansch equation, a set of hydrophobicity fragmental constants for the various amino acids and from these constants $\log P_{o,i}$, can be calculated using the equation

$$\log P_{o,i} = \sum a_n f_n \tag{15}$$

where a_n is a numerical factor indicating the incidence of a given fragment in a peptide structure and f_n represents the hydrophobic fragmental constant for each amino acid. Linear relationships are expected between $\log P_o$ (as calculated from hydrophobicity fragmental constant summations) and $\log k'$ (as determined with a particular alkyl silica–organic solvent–water system m) with slope $1/z_m$ for amphoteric molecules like peptides. Indeed, plots of $\log P_o$ versus $\log k'$ for the protein amino acids (*33, 34*) iodoamino acids (*35*), and families of peptide homologs (*30*) separated under isocratic elution conditions are essentialy linear, and become linear with gradient elution when corrections are made for the change in the volume fraction of the organic solvent modifier. Illustrative of this relationship is the plot of $\log P_o$ versus $\log k'$ for a series of iodotyrosines shown in Fig. 1.

In common with other polar solutes, peptide–nonpolar stationary phase interactions can be discussed in terms of a solvophobic model. In this treatment solute retention is considered to arise due to the exclusion of the solute molecules from a more polar mobile phase with concomitant adsorption to the hydrocarbonaceous bonded ligand, where they are held by relatively weak dispersion forces until an appropriate decrease in mobile-phase polarity occurs. This process can be regarded as being entropically driven and endothermic, i.e., both ΔS and ΔH are positive. Horvath and his co-workers have successfully adapted (*36*) the solvophobic theory as put forward by Sinanoglu *et al.* (*37, 38*) to evaluate $\Delta G^{\circ}_{\text{assoc}}$ and, hence, the factors affecting solute retention under a wide range of experimental conditions. This theoretical approach reveals that the capacity factor is a function *inter alia* of the interfacial surface tension

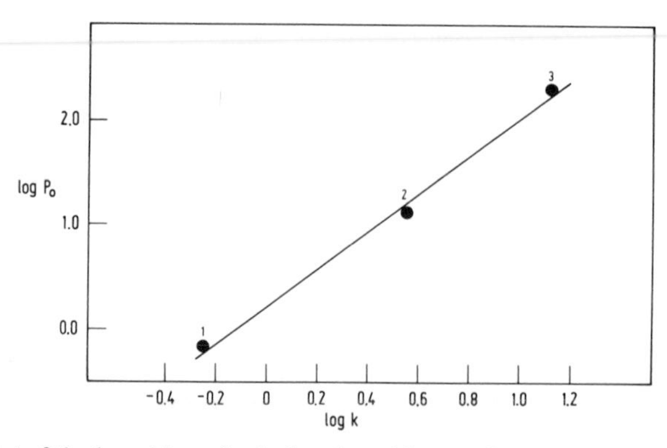

FIG. 1. Plot of the logarithm of calculated partition coefficients versus the logarithm of the capacity factors for tyrosine (1), 3-iodo-tyrosine (2), and 3,5-diiodotyrosine (3). The capacity factors were measured on a μ-Bondapak C_{18} column with a mobile phase of 10% methanol–water–5 m*M* orthophosphoric acid, adjusted to pH 3.0. Reprinted with permission from Hearn *et al.* (*10*). Copyright by Preston Publications Inc., Niles, Illinois.

γ, the static dipole moment μ of the solute molecule, and the relative surface area ΔA of the solute molecule in contact with the stationary phase, and can be indicated by the molecular surface area A, the polarizability of the solute, the static dielectric constant ϵ of the solvent, and the microscopic cavity factor κ^e, which may be defined as the ratio of the energy required to create a cavity for a solvent molecule to the energy required to extend the planar surface area of the solvent by the surface area of the added solute molecule. Since the interfacial tension between the mobile and stationary phases reflects the partitional properties of the two phases, changes in γ will have profound effects on peptide retention to nonpolar stationary phases. The relationship between retention and eluent surface tension may be given by

$$\log k = C + \gamma \frac{N\,\Delta A + 4.836N^{1/3}(\kappa^e - 1)V_{\mathrm{m}}^{2/3}}{RT} \tag{16}$$

where C is a constant, and N, V_{m}, R, and T are Avogadro's number, the average molar volume of the mobile phase, the gas constant, and the absolute temperature, respectively. Because water has the highest surface tension amongst all the common solvents, we can immediately recognize from Eq. (16) that maximal retention of peptides to nonpolar phases will occur with neat aqueous eluents. More importantly, theoretical treatments of this kind characterize the solvophobic strengths of mobile phases of different compositions (*17c, 30*). As will be outlined below, the experimentally observed behavior of peptides on reversed phases with aquo-organic solvent mobile phases is generally in accord with the relationships anticipated on the basis of solvophobic considerations. In particular, this approach permits an analysis of the effect of the organic solvent modifier, the ionic strength, surface-active ion-pairing reagents, ligand alkyl chain length, peptide composition and structure, and temperature on the sorptive behavior of peptides with hydrocarbonaceous stationary phases.

C. Stationary-Phase Effects

The presence of any residual silanol groups, which remain accessible to peptidic solutes after the alkyl bonding procedure, will result in mixed adsorption contributions in the interaction of peptides with the stationary phase. Polar adsorption effects between ionogenic groups on the peptide and the matrix may cause peak broadening, increased and erratic retention times, and poor resolution. For these reasons alone, chemically bonded alkyl silicas of high and uniform carbon loading of the support should be used for the reversed-phase HPLC separation of polar peptides. Although a relatively ordered, uniform monolayer of isolated alkyl "bristles" was originally envisaged for the ligand coat on the silica surface, there has been some recent discussion (*17a–17c, 39*) whether the bonded ligand may form energetically more favored solvophobic aggregates of nonpolar liquid droplet clusters with water-rich eluents but progressively solvate as the organic solvent content increases.

Because of steric restrictions, it is not possible to react all the free silanol groups in a porous silica matrix. With bonded reversed phases with carbon content above 10% (w/w), e.g., coverage ca. 1 mmol/g silica support, polar adsorption effects with unprotected peptides do not appear to give rise to deleterious loss of recoveries or poor peak shape. In fact, the involvement of polar adsorption processes can prove most advantageous as far as control over peptide selectivity on alky silicas is concerned. The chemical and physical characteristics of the parent silica matrix obviously will predetermine the retention features which can be expected for a reversed-phase sibling. Despite the exponentially expanding literature on reversed-phase HPLC of peptides, at this stage few truly systematic studies on the various available reversed-phase silicas (or for that matter, other chemically modified silica phases) of known different porosities, alkyl chain length or surface density, and silica matrix history have been completed. The naive researcher is, as a consequence, left with a large and at times bewildering array of different phases from which to choose and the choice once made may not necessarily be the most appropriate for the particular study. It is clearly unrealistic to expect any one reversed-phase silica to prove suitable for all separations and access to an array of different reversed-phase supports is essential for any laboratory contemplating the isolation of peptides from biological sources. Systematic characterization of alkyl silica phases will thus be essential if we are to take advantage of the full potential of HPLC methods for physicochemical measurements of, for example, peptide–ligand interactions or exploit the large data base potentially available for retention behavior prediction. In this context, it is interesting to note several studies (*39a–c*) addressed to the issue of the influence of alkyl chain length on peptide retention behavior. These studies have indicated that at constant surface coverage, little change in retention for polypeptides occurs. The inference can be drawn from this data that surface density rather than alkyl chain length is the significant controlling parameter. Although a number of short *n*-alkyl chain bonded supports have now been prepared by several groups specifically for the separation of nonpolar polypeptides, the claimed improvement in chromatographic performance may thus reflect the silica matrix and bonded carbon content rather than the alkyl chain length per se.

Nevertheless, the chemical composition of bonded ligand can affect selectivity and capacity factors of peptidic solutes. Thus, nonpolar peptides generally tend to show shorter retention times on the chemically bonded 3-cyanopropyl silicas and phenylpropyl silicas than on the more hydrophobic octadecyl silica. In fact the nonprotic bonded cyano phases can function as both normal and reversed phases of medium polarity and provide useful supports for the separation of hydrophobic peptides. The increased polarity of the bonded cyanoalkyl silicas can be mimicked with the *n*-alkyl silicas by the addition to the mobile phase of low concentrations of surface-active alcohols, e.g., *tert*-pentanol (*10, 40*). A general trend is apparent with peptides in which retention times but not selectivities are greater on C_{18} than on C_8 phases of similar coverage prepared from the same silica matrix. With the more modern octyl and octadecyl silicas, which have high carbon coverages and few accessible silanol groups, significant changes in the elution order for a series of simple nonpolar peptides rarely can be attributed solely to differences in the ligand chain length and more likely have their origin in the different adsorption isotherms due to the partition of the various components in mobile phases of differing composition. It is thus essential to ensure that a column is fully equilibrated to a particular mobile-phase condition, otherwise variability in column selectivity can arise. This is particularly important when the mobile phase contains surface-active buffer components such as sodium dodecyl sulfate (SDS). With some eluents, significant differences in resolution, retention times, and recovery for polar peptides have been noticed (*13, 27, 41, 42*) between the short chain, e.g., C_2 or the ethylphenyl, and the longer chain, e.g., C_8 or C_{18} ligands. Furthermore, with the more hydrophobic peptides some selectivity changes can be found between the capped and partially capped reversed phases, e.g., between Hypersil ODS and Spherisorb ODS (*8, 41*), indicative of polar adsorptive interactions.

Polypeptides with MW > 10,000 often show peak broadening on the 5- or 10-μm porous alkyl silicas now commonly in use for peptide separation. These supports have pore diameters typically on the order of 10 nm and the reduced efficiencies may be due to restricted diffusion of the polypeptide through the pores. The use of the larger-pore-sized silicas, e.g., 33, 50, or 100 nm, bonded with n-alkyl ligands has been found (*43, 44*) to circumvent, to some extent, the reduced resolution and improve recovery. Because polypeptides, proteins, and some peptides show very steep dependencies of their $\log k'$s on Ψ_0 (see Section III,E), the issue of optimal flow rate at any given eluent composition must, however, be taken into account with these larger-pore-size silicas.

Nonpolar, porous polystyrene–divinylbenzene (PS–DVB) copolymers have also been used (*41, 45a*) as reversed-phase HPLC packings for peptide separations. Compared to the chemically bonded n-alkyl silicas, these supports show generally much lower efficiencies. As they can, however, be used over a wide pH range, e.g., Amberlite XAD-4 can be used over ca. pH 1.5–12.5, these types of supports could find limited use in situations where pH control of the selectivity is mandatory.

D. Effect of Peptide Chain Length and Composition on Retention

There is now a general consensus that with reversed-phase separations of peptide mixtures with neat aqueous or aquo-organic solvent mobile phase of low pH, the capacity factors and elution order of a series of peptides reflect differences in the hydrophobicities of the solutes. Because of the very large variety of permutations possible with the protein amino acids, the structural diversity of peptides even of modest size is considerable. The polarity differences manifested by different peptides in solution will be dependent on a number of physicochemical factors and notably the organization arrangement of the amino acid side chains, the extent of ionization, and the involvement of secondary chemical equilibria in solution. Subtle modulation of these general effects can be achieved by changes in pH or with mobile phases containing low concentrations of counterionic reagents which undergo ion-pairing or dynamic liquid–liquid ion-exchange interactions.

Provided ionic adsorption and size-exclusion effects are kept minimal, the most polar peptide in a mixture will have the shortest chromatographic retention on nonpolar stationary phases with the remainder eluting in order of their relative hydrophobicities. This effect has been discussed (*17b, 39b, 45a*) in terms of the interfacial contact area for a peptidic solute–nonpolar stationary phase interaction. Briefly, this concept accommodates the composite effects which the compositional, sequential, and conformational parameters of a peptide have on retention, i.e., the additive hydrophobic influence of the different amino acid side chains substituted in an ordered manner into a polyglycinyl oligomer. In this treatment, molecular weight per se is not expected to dominate retention but rather the magnitude of the accessible hydrophobic contact areas. For small peptides, the relative hydrophobicities of the amino acid side chains would be expected to be the prevailing feature which influences retention, with the position and chirality of the side chains in relation to charged sites playing important, but secondary, roles. With larger peptides and polypeptides, sequential features, particularly if these result in the juxtapositioning of polar and/or hydrophobic domains, would be expected to become increasingly important in controlling retention under a particular set of elution conditions. A considerable body of chromatographic data is now in accord with these concepts.

Oligomers of glycine show little appreciable retention on octyl silica with 20 m*M* phosphate buffers over the pH range 2.10–7.83 (*46*), or on octadecyl silica with 100 m*M* phosphate buffer, pH 2.1 (*33*), or 5 m*M* phosphate buffers, pH 2.1, containing hydrophobic alkyl sulfonates (*30*). It is thus likely that the peptide chain proper makes only a very small contribution to the retention of peptides under these conditions. Based on partition coefficient considerations, oligomers of alanine, and the other nonpolar amino acids, should show a linear dependence of log k' on the number of residues. This, in fact, has been observed. For example, the plot of log k' versus the number of alanine residues shows (*33*) a linear dependence with a uniform log k' increment due to the methyl group of the aliphatic side chain (Fig. 2), i.e., the effect is additive (*45a, 46a*).

An interesting insight into the influence of charged groups, proximal to nonpolar residues, on peptide retention on reversed phases has been gained from studies with isomeric peptides. For example, DL-Leu-Gly will elute from a LiChrosorb RP8 before Gly-DL-Leu under acidic conditions (*46*) with an elution order reversal when the pH is increased to 7.90. This and other similar observations (*30*) suggest that with simple isomeric peptides, the position of the more nonpolar amino acid residues in relation to charged sites have a profound effect on the retention behavior. A general trend is apparent with isomeric peptides, for example, isomers of the type Gly-$(Y)_n$-X and X-$(Y)_n$-Gly, where X is a nonpolar or acidic amino acid, the isomer with the N-terminal nonpolar residue has the shorter retention on alkyl silicas under low-pH elution conditions. By

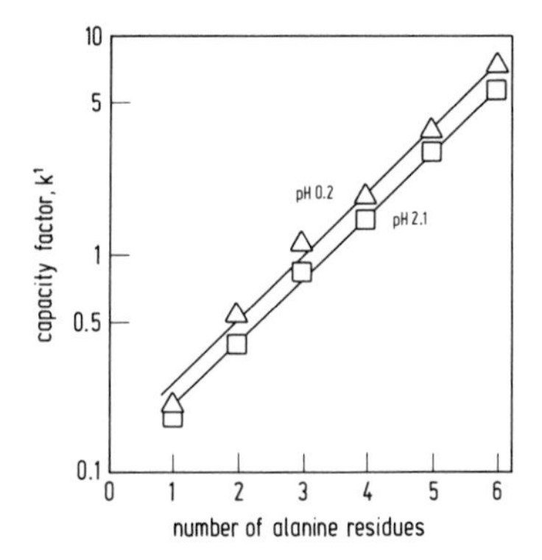

FIG. 2. Plot of the logarithm of the capacity factor against the number of residues in alanine oligomers. The capacity factors were obtained on a 5-μm LiChrosorb RP18 column with a 100 m*M* phosphate buffer; the flow rate was 2.0 ml/min. Reprinted with permission from Molnar and Horvath (*33*). Copyright by Elsevier Scientific Publishing Co., Amsterdam.

decreasing the surface exposure of the X-side chain moiety to the non-polar stationary phase, the cationic α-NH_3^+ has, in these cases, effectively reduced the hydrophobic contribution which the X-side chain can make to the retention phenomenon.

Replacement of a terminal hydrophobic group into an endo position of the peptide sequence can result in elution order changes for similar isomeric peptides. This effect, which can be attributed to a frame shift due to steric compression of the peptidic side chains leading to significant conformational changes, has been noted with a number of peptides (*40, 47–50*). For example, [βAla$_1$]-ACTH$_{1-18}$-NH$_2$ has a longer retention on Nucleosil C_{18} than [Gly$_1$]-ACTH$_{1-18}$-NH$_2$, but [βAla$_{10}$]-ACTH$_{1-18}$-NH$_2$ has a shorter retention than ACTH$_{1-18}$-NH$_2$ which has a glycine residue at position 10 (*50*).

Topological indices, based on hydrophobicity scales, have been used to predict the elution orders of small peptides (<20 residues) with reasonable accuracy. Molnar and Horvath found (*33*) a good correlation between the elution order of a number of small hydrophobic peptides, separated under gradient elution conditions, and the hydrophobic fragmental constants as derived by Rekker. This approach has been extended by Hearn *et al.* (*30, 35*) for the prediction of the elution orders for iodoamino acids and polar peptides and by O'Hare and Nice (*17*) for hormonal peptides. Under isocratic elution conditions a good correlation with k' versus fragmental constant summation can be obtained with simple peptides with the slopes of the straight lines corresponding to the elutropic strength parameter z_m, as defined by Eq. (*14*). When the change in elutropic strength Δz_n for gradient elution is taken into account, similar types of correlation are observed, i.e., the adjusted k' of a series of phenylalanine oligomers are related to a linear function of the number of residues (*45a, 46a*). Analogous calculations based on the hydrophobicity scales of Nozaki and Tanford (*51*) and Hansch *et al.* (*52*) have been found to be less applicable to retention order predictions for unprotected peptides separated on alkyl silicas. Amino acid retention parameters for the separation of peptides in perchlorate systems have been recently described (*46a*).

Based on the theoretical arguments summarized in Eqs. (12) and (16), selectivity differences can be evaluated in terms of group contribution coefficients through the dependence of $\log \alpha_{i,j}$ [$= \log k'_i/k'_j = \Delta(\Delta G^\circ_{\mathrm{assoc},i,j})/RT$] on changes in the hydrophobic contact area, $\Delta\Delta A_{i,j}$. At this stage there is no general *ab initio* method to derive ΔA values of a group of peptides for a particular chromatographic system or to relate conformational and structural determinants as revealed by analysis of X-ray crystallographic data (*52a*) to retention behavior. With homologous peptides, plots of $\log k'$ versus the number of residues have been found experimentally to be essentially linear, i.e., $\Delta\Delta A$ remains constant. Interestingly, although $\log k'$ values show uniform incremental changes (as determined by peak moment measurements) for a homologous peptide series up to ca. $n = 10$, peak efficiencies become dramatically reduced with the larger oligomers. This effect is probably due to low solute solubility associated with poor desorption kinetics. Oligophenylalanines, oligoglutamic acids, and oligolysine derivatives can be used to evaluate alkyl silica performance and characteristics (*39b*) based on an analysis of the $\log k'$ versus peptide chain length relationship.

In order to be able to predict the retention behavior of peptides of different composition, of peptides of the same composition but different sequence (positional isomers), and of diastereoisomeric peptides, a knowledge of the incremental contribution of each amino acid to the overall contact area term is required not only for each well-defined stationary phase but also for each mobile-phase condition. Group retention coefficient summation approaches based on the assumption that selectivity differences can be ascribed predominantly to amino acid sequence differences, have been developed by Meek (*46a, 52b*) and Su *et al.* (*45a*). These treatments have subsequently been applied to a number of different elution systems (*52c–52e*). A comparative analysis of the different amino acid group contribution coefficients derived for phosphate, perchlorate, pyridine/acetate, trifluoroacetate, and bicarbonate buffer systems has been reported (*52f*).

Studies with larger peptides, e.g., polypeptide hormones, have indicated that the resolving power of reversed-phase separations may decrease as the peptide chain length increases. These results stress the importance of using elution conditions which generate adequate relative selectivity factors ($\Delta\alpha/\alpha$). As can be seen from Eq. (1), a 10-fold decrease in the relative selectivity factor would require a 100-fold increase in theoretical plate numbers to offset the loss in resolution. For larger peptides (>20 residues) decreased selectivities are anticipated with unbuffered aquo-organic solvent eluents. This situation is paralleled by a diminished predictive value of the topological fragmental constant summations for polypeptides. This is not unexpected since these parameters consider only the compositional, and not secondary or tertiary structural, value of the amino acid side chains. Both these structural effects will reduce the number of exposed hydrophobic residues and lead to elution order anomalies based solely on the sum of the side-chain hydrophobicity terms. Even with relatively small polypeptides, e.g., β-endorphin analogs, in which stable secondary structures are not anticipated, retention anomalies can be anticipated. Despite these discrepancies, the current summation procedures provide a useful tool for elution order predictions but point to the need for hydrophobicity tables which account for secondary structural effects.

E. Mobile-Phase Effects Involving Organic Solvent Modifiers

Because the potential for variation in both composition and structure of natural and synthetic peptides is very large, elution conditions must be able to take advantage of those secondary chemical equilibria which are responsive to polarity differences expressed by the solutes in solution. Neat aqueous eluents alone will not exhibit a sufficient range of elutropic strengths, irrespective of the chosen pH or ionic strength, to elute, from chemically bonded reversed phases with realistic k', peptides differing even modestly in polarity. To achieve a wider range of eluent strengths, an alteration in the bulk structure of the eluent is required and this can most easily be accomplished by the use of organic solvent modifiers. This will result in a reduction in surface tension γ of the mobile phase that is discrete with isocratic elution and continuous with gradient elution. Since $\log k'$ is linearly related to γ [see Eq. (16)], it should be possible to estimate the k' values for peptides under different mobile-phase conditions using γ values (*30*). The lack of sufficient literature γ values for multicomponent mobile phases limits this approach at present, although empirically elution conditions can be chosen by comparing the $\log k'$ trends as the relative percentages of two or more organic solvent modifiers are changed (*30, 40*).

Methanol, acetonitrile, and 1- or 2-propanol have been the most popularly used solvents for peptide separations, although other water-miscible, UV-transparent solvents (e.g., methoxyethanol, ethanol, butanol, tetrahydrofuran, and dioxane) have also been employed. The retention of peptide solutes generally follows an inverse relationship to the elutropic value of these common organic solvents. Of practical significance is the fact that some selectivity factor changes are expected on the basis of solvophobic and conformational arguments when one solvent is changed to another solvent. A number of examples of specific solvent-dependent elution order reversals have been reported (*17, 47, 53*). These selectivity effects are probably due to solvent-induced conformational changes in the peptides and these will lead to changes in the surface hydrophobicities, the magnitude of which concomitantly depends on the nature of the solvent and the solute. Because of their intermediate polarity [$\epsilon^{\circ}(Al_2O_3) = 0.82$], 1- and 2-propanols offer some selectivity advantages, despite their high viscosity ($\eta = 2.37$), for the separation of very hydrophobic peptides or large polypeptides. There is, however, an apparent decrease in column efficiency for peptide separations with these propyl alcohols compared to acetonitrile [$\epsilon^{\circ}(Al_2O_3) = 0.64$, $\eta = 0.37$] and this suggests that, in general, the reduced plate height h will be dependent on eluent viscosity.

It has been concluded (*54*) on the basis of theoretical arguments that $\ln k'$ for small solutes varies quadratically with the volume faction ψ of the organic solvent modifier in a binary mixture, i.e.,

$$\log k' = A\psi^2 + B\psi + C \tag{17}$$

and this has been observed experimentally for a variety of benzene derivatives. Peptides also show a nonlinear dependence of k' on ψ, with k' falling off rapidly with increasing concentration of the organic solvent modifier up to ca. $\psi < 0.5$. Because of this dependency, significant variations in retention can be observed with peptides under isocratic conditions which may differ only over a narrow range of organic solvent modifier concentrations, i.e., by a few percent.

In detailed studies examining the influence of the volume fraction of the organic modifier, ψ, on k' Hearn and Grego have shown (*17a–17c, 30, 44, 54a*) for a large variety of peptides, polypeptides, and proteins that bimodal dependencies exist between ψ and $\log k'$ over the full $0 \leq \psi \leq 1$ range for methanol, acetonitrile, and the propanols. Regular reversed-phase behavior is observed with water-rich eluents but above $\psi > 0.5$ elution order reversals can occur reminiscent of a normal-phase elution mode. Furthermore, with increasing peptide chain length and hydrophobicity the dependencies between $\log k'$ and ψ become progressively more severe. Little or no elution development is evident for larger, nonpolar polypeptides. Typical of these observations are the data shown in Figs. 3 and 4 for several polypeptide hormones and phenylalanine oligomers where the elution order reversals are particularly striking. It has been argued that the increases in k' values seen with polar solutes on alkyl silicas at high organic solvent concentrations is due to silanophilic interactions (*54b*)

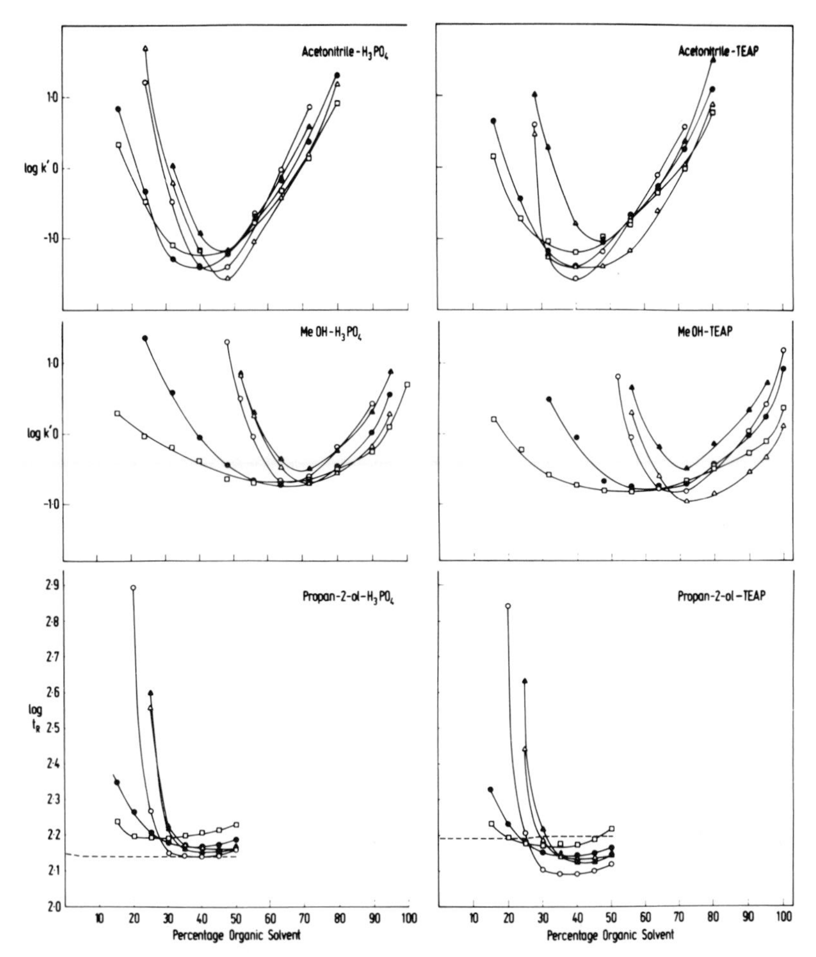

FIG. 3. Plots of the logarithmic retention factors of several polypeptide hormones against the volume fraction of organic solvent in the aquo-organic solvent mixture used as the eluent. A, C, and E— data were obtained with the organic solvent modifiers acetonitrile, methanol, and propan-2-ol, respectively, with a primary mobile phase comprised of water–20 m*M* orthophosphoric acid. B, D, and F— data were obtained with the organic solvent modifiers acetonitrile, methanol, and propan-2-ol, respectively, using a primary mobile phase comprised of water–15 m*M* triethylammonium phosphate. The dashed lines in E and F represent the logarithmic retention times for sodium nitrate. Column: μ-Bondapak C_{18}; flow rate: 2.0 ml/min for the acetonitrile and methanol experiments, 1.2 ml/min for the propan-2-ol experiments. The polypeptide key is as follows: ●, angiotensin 1; □, angiotensin 11; ○, bovine insulin; △, bovine insulin B chain; ▲, porcine glucagon. Reprinted with permission from Hearn and Grego (*17b*). Copyright by Elsevier Scientific Publishing Co., Amsterdam.

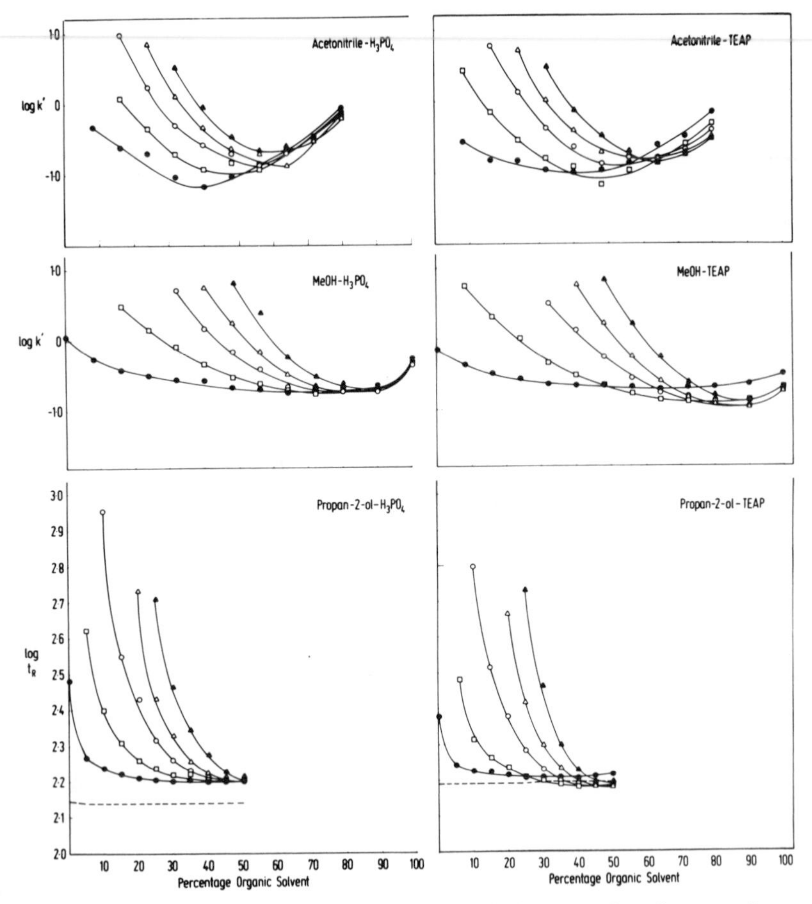

FIG. 4. Graphs illustrating the relationship of the retention factors of several phenylalanine oligomers on octadecyl silica and the composition of the aquo-organic eluent. The phenylalanine oligomer key is as follows: ●, F; □, FF: ○, FFF; △, FFFF; ▲, FFFFF. The chromatographic conditions are itemized in the legend to Fig. 3.

between the solute and the silica matrix. Since these effects can only be partially abolished by the use of alkylammonium buffers in the mobile phase, further explanations must be sought. In this context the solvent-dependent selectivity changes seen with peptides may provide a clue. Such subtle processes are indicative of hydrogen bonding and solvation phenomena, and these effects, when taken in association with the known solvent distribution which occurs at the hydrocarbonaceous stationary phase–mobile phase interface, may provide a possible basis for the observed decrease in resolving power seen with polypeptides that elute in the range $0.35 < \psi < 0.55$ (*30, 39a,b*).

Gradient elution has proved essential for the separation of complex mixtures of peptides. Because each peptide exhibits unique $\log k'$ versus ψ dependencies and solubility characteristics, attention must be given to the choice of optimal gradient shape and flow rate. To obtain efficient chromatography under gradient conditions, it is essential to use a slow rate of change of the organic solvent modifier, with 0.2–0.5%/min at a flow rate of 2.0 ml/min being a common choice. It has been the author's experience that linear gradient shapes with acetonitrile give better resolution for complex mixtures of peptides than do exponential shapes, whereas the opposite situation generally holds for methanol and the other alcohols. Figure 5 shows a typical example of the influence of gradient shape on resolution, the sample in this case being the tryptic digest of hemoglobin A; the mobile phases were generated under three different gradient shape conditions using a 0–50% acetonitrile–water–0.1% orthophosphoric acid system (*55*). The above conclusions on optimal gradient shapes for peptide separations with a particular solvent, parallel similar observations made for small organic solutes. Ideally, with shallow gradients, i.e., when $d\psi/dt$ is small, and with relatively high flow rates it should be possible to achieve optimal resolution and recovery. This, in fact, has been observed (*17c, 39b*) for both peptides and proteins eluted under linear solvent strength (LSS) gradient conditions.

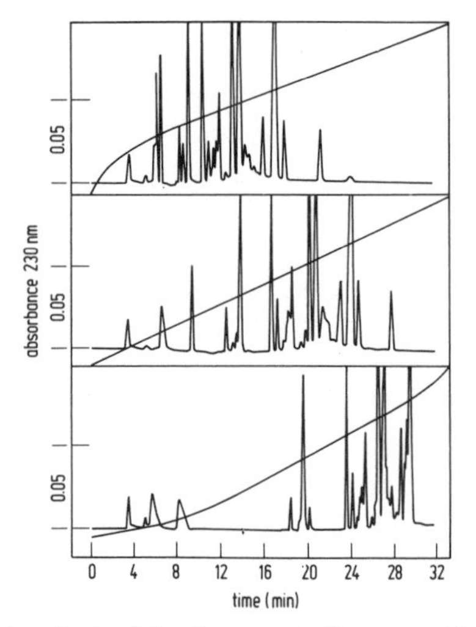

FIG. 5. Elution profiles obtained for the tryptic digest peptides of hemoglobin A using three distinct gradient shapes (number 3, 6, and 8 on the Waters M660 solvent programmer). In each case, two μ-Bondapak C_{18} columns were connected in series, a flow of 1.7 ml/min was used, and a 30-min linear gradient was generated from solvent A (0.1% phosphoric acid, pH 2.2) and solvent B (50% acetonitrile–0.1% phosphoric acid, pH 2.2). Reprinted with permission from Bishop *et al.* (*55*). Copyright by Marcel Dekker, Inc., New York.

Determinations of the adsorption isotherms for a number of organic solvent–water systems in contact with hydrocarbonaceous stationary phases have shown that a layer of solvent molecules forms on the bonded-phase surface and that the extent of the layer increases with the concentration of the solvent in the mobile phase. For example, methanol shows a Langmuir-type isotherm when distributed between water and Partisil ODS (*56*). This effect can be exploited to enhance the resolution and the recoveries of hydrophobic peptides by the use of low concentrations, i.e., $<5\%$ v/v, of medium-chain alkyl alcohols such as *tert*-butanol or *tert*-pentanol or other polar, but nonionic solvents added to aquo-methanol or acetonitrile eluents. It also highlights the cautionary requirement that adequate equilibration of a reversed-phase system is mandatory if reproducible chromatography is to be obtained with surface-active components in the mobile phase.

From a practical point of view, the organic solvent modifier must be chemically and physically compatible with the solute, the materials used in the column supports, solvent delivery systems, and detectors. When a choice exists, mobile phases of lower viscosity are preferable since they can be used over a wider range of flow velocities without generating extremely high backpressures. For detection convenience, the organic solvents used should be UV-transparent, preferably down to 200 nm. At this wavelength, UV spectrophotometers function as universal detectors and reveal the resolved solutes as well as the presence of eluent artifacts. Although this solvent background information may not be required and can be tedious, as, for example, when baseline changes occur in gradient elution, UV detection near the isosbectic maxima of peptides is often preferable and is essential when the peptide lacks an aromatic chromophore. Baseline adjustment for solvent artifacts, and solute peak amplification, can be readily carried out with a variety of laboratory automation computers such as the Hewlett-Packard model 3354. Fluorometric detection, either via endogenous tryptophan fluorescence or following postcolumn derivatization with a fluorochrome, is a powerful adjunct method to monitor eluents. The use of fluorometric detection of peptides in reversed-phase HPLC has recently been reviewed (*57*). With suitable mobile phases absorption ratioing at different wavelengths can be used as a powerful probe with both UV and fluorometric detectors to indicate or confirm molecular features.

F. Effect of pH on Peptide Retention

Peptides, being ionogenic substances, undergo protic equilibria. This property has long been exploited for their separation by the classical ion-exchange and liquid–liquid partition chromatographic procedures. It is an important parameter in regulating solute retention with chemically bonded reversed phases. Since retention is dependent on the solute's polarity, ionization of free amino or carboxyl groups to their respective conjugate acidic or basic forms will increase the polarity of a peptide, and hence will reduce its retention. Simple peptides with no ionogenic side chains will ionize according to

$$\overset{+}{H_3N}\overset{R}{\overset{|}{C}}HCO{-}(NH\overset{R^x}{\overset{|}{C}}HCO)_n{-}NH\overset{R'}{\overset{|}{C}}HCO_2H + H_2O \underset{}{\overset{K_{a1}}{\rightleftharpoons}} H_2N\overset{R}{\overset{|}{C}}HCO{-}(NH\overset{R^x}{\overset{|}{C}}HCO)_n{-}NH\overset{R'}{\overset{|}{C}}HCO_2H + H_3\overset{+}{O}$$

$$H_2N\overset{R}{\overset{|}{C}}HCO{-}(NH\overset{R^x}{\overset{|}{C}}HCO)_n{-}NH\overset{R'}{\overset{|}{C}}HCO_2H + H_2O \overset{K_{a2}}{\rightleftharpoons} H_2N\overset{R}{\overset{|}{C}}HCO{-}(NH\overset{R^x}{\overset{|}{C}}HCO)_n{-}NH\overset{R'}{\overset{|}{C}}HCO_2^- + H_3\overset{+}{O}$$

Although the side-chain groups R, R^x, R′ have only a weak influence on the acid dissociation constants K_{a1} and K_{a2}, the reverse situation, namely a change in polarity due to an ionization event in the vicinity of R, R^x, R′, can have a significant influence on retention. The capacity factor equation describing the sorption to a reversed phase of the above simple nonpolar peptide as a function of pH is given by

$$k' = \frac{k_0 + k_1([H^+]/K_{a1}) + k_2(K_{a2}/[H^+])}{1 + [H^+]/K_{a1} + K_{a2}/[H^+]} \tag{18}$$

where k_0, k_1, and k_2 are the capacity factors of the zwitterionic, cationic, and anionic species, respectively. Since in the region of the isoelectric point both the amino and carboxyl groups will be ionized (zwitterionic form), retention will be low. Theory thus predicts that the retention of peptides to reversed phases will progressively increase as the pH is lowered below the isoelectric point. Kroeff and Pietrzyk (*45*) have observed this effect for the tetrapeptide (L-Ala)$_4$ and the dipeptide L-Ala-L-Leu on the nonpolar, porous Amberlite XAD-4 and XAD-7 supports, respectively, whereas Hearn *et al.* (*29*) have reported a similar dependency for X-Tyr dipeptides on μ-Bondapak C_{18}. Typical of this dependency is the peptide k' versus pH data on μ-Bondapak C_{18} shown in Fig. 6. For peptides with side-chain ionogenic groups, as well as polypeptides in general, the capacity factor dependency on pH can be written

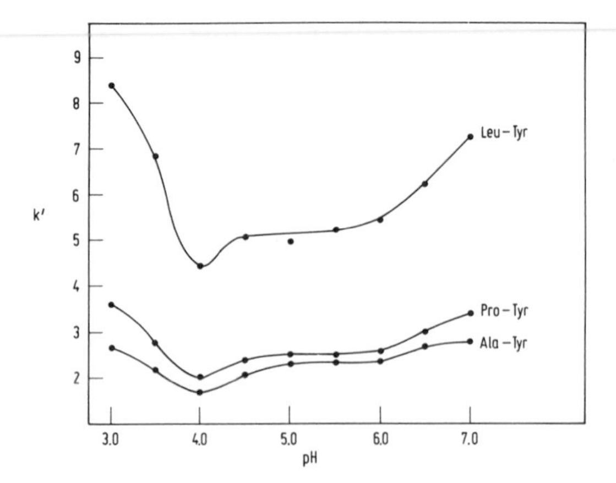

FIG. 6. Capacity factors for the X-Tyr peptides on a μ-Bondapak C_{18} reversed-phase column as a function of pH. The eluent was 5% methanol–95% water–5 m*M* H_3PO_4, titrated to the appropriate pH with NaOH, at a flow rate of 2 ml/min. Reprinted with permission from Hearn *et al.* (*29*). Copyright by Elsevier Scientific Publishing Co., Amsterdam.

in terms of an expanded polynomial expression which for acidic or tyrosinyl peptides takes the form

$$k' = \left[k_0 + k_1\left(\frac{[\mathrm{H}^+]}{K_{\mathrm{a1}}}\right) + k_2\left(\frac{K_{\mathrm{a2}}}{[\mathrm{H}^+]}\right) + k_3\left(\frac{K_{\mathrm{a2}}K_{\mathrm{a3}}}{[\mathrm{H}^+]^2}\right) + \cdots + k_n\left(\frac{K_{\mathrm{a2}}K_{\mathrm{a3}}\cdots K_{\mathrm{a}n}}{[\mathrm{H}^+]^n}\right)\right] \times \left[1 + \frac{[\mathrm{H}^+]}{K_{\mathrm{a1}}} + \frac{K_{\mathrm{a2}}}{[\mathrm{H}^+]} + \frac{K_{\mathrm{a2}}K_{\mathrm{a3}}}{[\mathrm{H}^+]^2} + \cdots + \frac{K_{\mathrm{a2}}K_{\mathrm{a3}}\cdots K_{\mathrm{a}n}}{[\mathrm{H}^+]^n}\right]^{-1} \quad (19)$$

where $K_{a1}, K_{a2}, K_{a3}, \ldots, K_{an}$ are the ionization constants for the n-step ionization process and $k_0, k_1, k_2, k_3, \ldots, k_n$ are the capacity factors for the zwitterion, monoprotonated species, singly charged anionic species, doubly charged anionic species, etc. Equations of this type can be used to provide a qualitative description of the pH-dependent equilibria that influence the retention of peptides on reversed phases. For example, it is possible to determine with this method K_a values for a peptide since k' data can be readily determined as a function of pH. Most importantly in a practical sense, elution conditions can be chosen on the basis of ionization and pH considerations when the compositional features of a peptide are known. Alternatively, they may be predicted from k' versus pH plots, which can be calculated from equations of the above type using a minimum number of experimentally derived k' values. Such a treatment allows pH versus resolution optima to be chosen for mixtures of peptides using a window diagram approach similar to that originally proposed for the optimization of the separation of reversed-phase HPLC of weak organic acids (*58*).

With water–organic solvent mixtures, the extent of ionization of a particular peptide will vary according to the solvent use due to the participation of specific solvation effects and to a nonlinear dependency of protonic equilibria on the dielectric constant of the mobile phase. Since the extent of retention to an alkyl-silica support is strongly influenced by the ionization state of the peptide, solvent-dependent, pH-mediated effects will give rise to significant selectivity changes. It is noteworthy that these pH effects are manifested with chemically modified nonpolar silicas and the Amberlite XAD-type resins. For example, the elution order reversal of L-Phe-Gly and Gly-LD-Phe on Amberlite XAD-2 has been observed (*45*) when the pH was changed from 2.73 to 11.03, and similar effects have been noted with amino acids and peptides chromatographed on n-alkyl silicas (*42, 46, 47, 59*). Because of the chemical instability of n-alkyl silicas at high pHs, most separations on these supports use mobile phases with pH < 7.5 and usually in the range pH 2–3. At these low-pH conditions, which are below the isoelectric points of most peptides, ionic suppression of the carboxyl groups will occur and the percentage protonations of the amino groups will generally be greater than 99.9999. Consequently, increased solute retention and improved peak shape is expected under these conditions. Furthermore overall decreases in column selectivities have been noted (*29, 46, 60*) for the separation of peptides on both octyl and octadecyl silicas as the pH is raised from 3.0 to 7.0. With tyrosinyl peptides, such as the oxytocins, a progressive increase in k', but decrease in α value, has been observed (*47*) in the range pH 5–8, followed by a fall in k' at higher pH. The latter decrease in k' may be attributable to the ionization of the phenolic hydroxyl group of the tyrosinyl side chain. The effect of low pH (e.g., pH 2.5) on a peptide containing a lysinyl, hydroxylysinyl, histidinyl, or argininyl residue will be to further reduce its k' compared to, say, the corresponding isoleucyl analog due to the formation of cationic charged centers via protonation of the ϵ-amino, imidazolyl, or guanidino side-chain groups. Conversely, peptides with aspartyl or glutamyl residues will become less polar due to the low-pH conditions and their k' will increase. Peptides with acidic residues, and to a lesser extent those with asparaginyl or glutaminyl residues, show reduced retention to reversed phases at higher pH. One possible disadvantage of low-pH conditions with the silica-based bonded reverse phases is the potential for adsorption of the protonated peptide to accessible silanol groups on the stationary phase. In practice, this is not a major problem and excellent recoveries can generally be obtained. If difficulties do arise with refractory peptides under these low-pH conditions, these can usually be circumvented by the addition of suitable buffer counterions, at appropriate ionic strengths, to the mobile phase. Alternatively, stationary phase-modifying reagents, which may be either ionic or nonionic, can be used.

G. Effect of Buffer Interactions on Peptide Retention

Variation of the pH or water content of a mobile phase may not necessarily permit adequate column selectivities for peptide separations, e.g., basic hydrophilic peptides may be insufficiently retained on reversed phases at low pH. By increasing the ionic strength, the surface tension of water–organic solvent mixtures is also increased and advantage can then be taken of the "salting out" effect. Based on the theoretical considerations summarized by Eq. (16), a rise in k' values for peptides would be expected with increasing ionic strength. This has been observed with simple eluents for both acids and bases in their ionized forms and particularly with strong organic acids like the sulfonates for which both k' and column efficiency increase with increasing ionic strength (*61*). However, the reverse trend is found (*31, 62*) with these types of solutes when the mobile phase contains ion-pairing reagents. In the absence of pH control with simple aqueous eluents, the k' values of amino acids initially decrease but then increase as the salt (e.g., NaCl) concentration is increased. The major salt effect noted with buffered aquo-organic solvent mixtures and, in particular, with phosphate systems, for peptide separations on reversed phases, is a decrease in k' values as the molarity of the buffer is increased (*30, 42, 45, 47*). At high ionic strengths, peptide recoveries from reversed-phase silicas are often impaired. These effects have been discussed in terms of solute–buffer ion interactions (*30, 47*).

In a series of studies we recently demonstrated (*29, 30, 63–67*) that the resolution of peptides on reversed phase can be profoundly influenced by the addition of appropriate counterionic reagents to a mobile phase of defined pH, ionic strength, and water content. Retention under these conditions can be discussed in terms of ion-air associations between the ionized peptide and a counterion in the mobile phase and subsequent sorption of the complex onto the stationary phase. Alternatively, adsorption of the counterion, particularly if it is lipophilic, onto the nonpolar stationary phase may occur, and peptide retention would then be mediated by dynamic liquid–liquid ion-exchange effects. Arguments in favor of the participation of one, the other, or both of these alternative pairing-ion phenomena in "ion-pair" chromatography have been extensively reviewed (*16, 28b, 62, 68, 68a*). It can be shown (*62, 68*) that retention behavior in ion-pair systems can be described by

$$k' = (k_0 + \beta[X])(1 + K_1[X])^{-1}(1 + K_2[X])^{-1} \qquad (20)$$

where k_0 is the solute capacity factor in the absence of the pairing ion X; [X] is the pairing-ion concentration; K_1 is the ion-pair formation constant; K_2 is the pairing ion–ligand binding constant; and β depends on the underlying physicochemical equilibria which control retention, i.e., whether ion-pair adsorption or dynamic ion-exchange mechanisms predominate. Provided $1/K_2[X] < 1/K_1$, a plot of k' versus [X] according to Eq. (20) yields a parabola, whereas in circumstances where the direct binding of the pairing ion to the nonpolar ligand is not significant, i.e., $K_2[X] << 1$, the dependency of k' on the pairing-ion concentration is hyperbolic. With amino acids and simple peptides, both types of dependency have been observed (*29, 30, 69*). Typical of these results are the plots of k' versus [hexylsulfonate] for several basic peptides shown in Fig. 7. Programs to permit nonlinear least squares analysis of retention data for peptides chromatographed or alkyl silicas in the presence of varying concentrations of pairing ions have been published (*70*).

Since pH conditions can be readily chosen to ensure that peptides are significantly ionized, peptide elution orders from nonpolar phases with ion-pairing elution systems will be determined by their relative hydrophobicities and the polarity or charge density of the counterion. Major changes in selectivity and retention times of peptides on reversed-phase

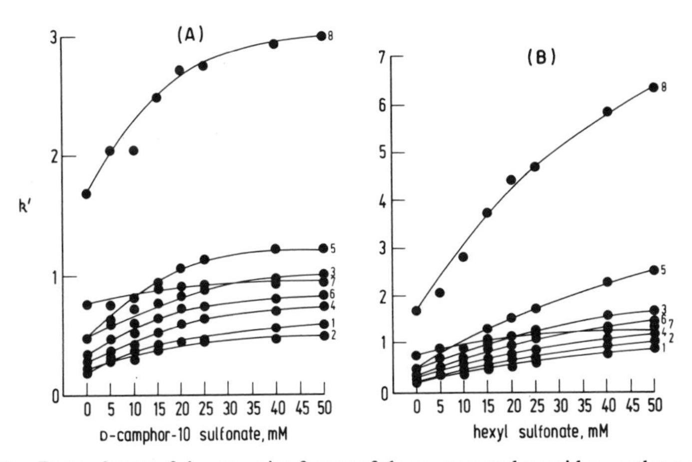

FIG. 7. Dependence of the capacity factor of the protonated peptides on the concentration of D-camphor-10 sulphonate (A) and *n*-hexyl sulfonate (B) in the mobile phase. Column: μ-Bondapak C_{18}; flow rate: 2 ml/min; temperature: 20°C, mobile phase: 50% methanol–50% water–50 m*M* NaH_2PO_4 with H_3PO_4 added to pH 3.0, containing various concentrations of the ion-pairing reagents. The protonated peptides were as follows: 1, Arg-Phe; 2, Arg-Phe-Ala; 3, Met-Arg-Phe; 4, Met-Arg-Phe-Ala; 5, Leu-Trp; 6, Leu-Trp-Met-Arg; 7, Leu-Trp-Met; 8, Leu-Trp-Met-Arg-Phe. Reproduced from Hearn and Grego (*34*).

systems can thus be achieved by judicious changes in the chemical nature of the counterion. For example, a peptide containing an N-terminal basic amino acid (e.g., lysine) will have at low pH a shorter retention to octadecyl silica when a hydrophilic anionic pairing ion is present in the mobile phase than an analog with, say, an N-terminal leucine. When a hydrophobic anionic or cationic pairing ion (e.g., butane sulfonate, tetramethylammonium salts) is present in the mobile phase at a particular pH, both k' and α changes are expected due to solute–pairing ion interactions. With the more strongly lipophilic anionic or cationic reagents (e.g., dodecyl sulfate or dodecylammonium salts) major selectivity changes will arise with polar peptides due to the generation of dynamic liquid–liquid ion exchangers. Thus at low pH, hydrophobic anionic reagents will generally increase retention, whereas the reverse will occur with cationic reagents. Furthermore, more subtle selectivity effects will become evident under pH-controlled conditions. For example, at pH 2.0, the cationic counterions will allow greater control over selectivity for peptides differing only in the residues Glu versus Gln or Asp versus Asn. Illustrative of these points are the selectivity plots shown in Fig. 8 for several peptides separated on an octadecyl-silica support in the presence of a fixed concentration of pairing ion (hexylsulfonate) but at different pHs. Although some differences of interpretation concerning the mechanism of action of surface-active pairing ions still remain (*56*), the application of ion pair–dynamic ion exchanger concepts to the reversed-phase HPLC separation of peptides has allowed many of the retention effects arising from the addition of ionic components to the mobile phase to be rationalized. More importantly, the practical usage of these concepts has considerably widened the scope of the technique. The use of pairing-ion elution systems, in fact, has revolutionized the analysis and separation of peptides from biological sources since it is possible to carry out direct radioimmunoassays for specific peptides in the presence of many types of pairing ions, e.g., the hydrophobic alkylsulfonates. Furthermore, paired-ion complexes of peptides show higher volatilities and this property can be exploited in on-line LC–MS analysis. Table I illustrates the retention effects which can be readily achieved by the simple expedient of changing the pairing ion and pH for a series of argininyl peptides eluted under isocratic conditions. A large number of anionic and cationic reagents have now been applied to analytical separations of peptides on reversed phases (see Refs. *66–68* for introductory compendia on the theoretical and practical aspects of these applications). A selection of the more common ionic and neutral species which modify the retention of peptides on reversed phases is found in Table II.

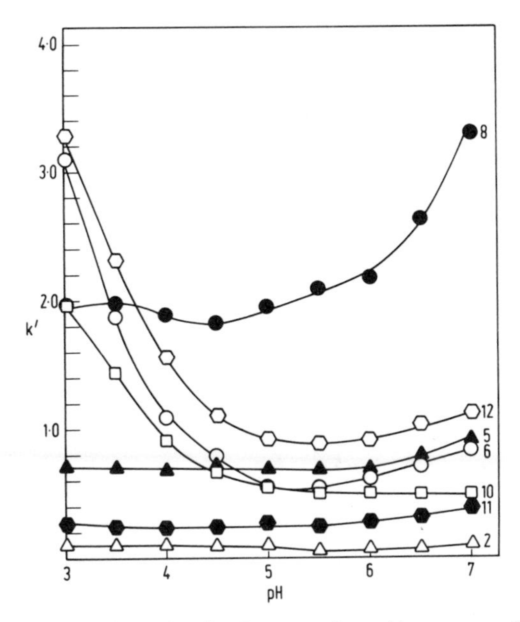

FIG. 8. Capacity factor dependencies for several peptides separated on a μ-Bondapak C_{18} column as a function of pH in the presence of a fixed concentration of pairing ion reagent. The eluent was 20% methanol–80% water–50 mM NaH_2PO_4–15 mM H_3PO_4–10 mM sodium hexane sulfonate, titrated with 10 M NaOH over the range pH 2.5–7.0; flow rate: 1 ml/min. The peptides were as follows: 2, G.G.; 5, G.F. amide; 6, R.F.; 8, F.L. amide; 10, G.L.Y.; 11, G.G.Y.; 12, R.F.A. The one-letter code is A, Ala; G, Gly; F, Phe; L, Leu; Y, Tyr; R, Arg.

The use of ammonium, alkylated ammonium, pyridinium, morpholinium, and a variety of inorganic cationic salts has been variously reported for peptide separations by reversed-phase HPLC. Besides participating in ionic equilibria with the peptidic solutes, these cations can interact with any accessible silanol groups and thus reduce polar adsorption effects. However, the capacity factor changes which have been noted (*65*) for peptides in the presence of tetra-*n*-alkylammonium and *N*-alkylammonium salts suggest that the dominant effects mediated by these cations have their origin in ion-pair–dynamic ion-exchange mechanisms rather than silanol inactivation. There is also a clear trend with the alkylammonium salts (and a corresponding situation is evident with the anionic alkylsulfonates and alkylsulfates) that as the hydrophobicity of the ion increases, the capacity factors of the peptidic solutes also increase without affecting the elution order. Illustrative of the use of an alkylammonium system is the separation (*88*) of bovine despenta insulin from

TABLE I

Capacity Factors for Various Small Peptides with Different Pairing Ion Systems

Peptide[b]	Capacity factors[a] k					
	(a) $H_2PO_4^-$	(b) Hexyl sulfonate	(c) Camphor-10-sulfonate	(d) Dodecyl sulfate	(e) Dodecylammonium phosphate	(f) Tetrapropylammonium phosphate
RF	0.20	0.46	0.37	8.8	0.18	0.22
RFA	0.22	0.37	0.30	8.2	0.18	0.24
MRF	0.48	0.71	0.60	>16	0.32	0.22
MRFA	0.28	0.58	0.42	13.0	0.22	0.22
LW	0.48	0.70	0.72	N.D.[c]	0.20	0.25
LWM	0.77	0.92	0.81	N.D.	0.18	0.28
LWMR	0.34	0.69	0.52	3.8	0.35	0.60
LWMRF	1.69	2.81	2.04	11.0	0.90	1.40

[a] Column: μ-Bondapak alkylphenyl using methanol–water (50 : 50): (a) 65 mM, pH 3.0; (b) 10 mM, pH 3.0; (c) 10 mM, pH 3.0; (d) 5 mM, pH 7.2; (e) 2 mM, pH 4.0; (f) 2 mM, pH 4.0.

[b] The one letter code for the amino acids is used as proposed by M. O. Dayhoff *in* "Atlas of Protein Sequence and Structure," National Biomedical Research Foundation, Silver Spring, Maryland: A, alanine; F, phenylalanine; L, leucine; M, methionine; R, arginine; W, tryptophan.

[c] N.D., Not determined.

TABLE II

Selection of Ionic and Neutral Species Which at pH < 7.0 Modify the Retention Characteristics of Unprotected Peptides on Chemically Bonded Hydrocarbonaceous Stationary Phases[a]

Anionic pairing ion	Reference[b]	Cationic pairing ion	Reference[b]
Ionic-Pairing Ion/Dynamic Liquid–Liquid Ion Exchange/Mobile-Phase Buffer Species			
$H_2PO_4^-/HPO_4^{2-}$	*8, 10, 24, 33, 46, 47, 60, 63, 66, 67, 71–75*	NH_4^+	*49, 60, 65, 80, 87*
Cl^-	*17, 39a, 45, 71–75*	$CH_3\overset{+}{N}H_3$	*65*
ClO_4^-	*33, 46a, 66*	$C_3H_9\overset{+}{N}H_3$	*65*
SO_4^-	*66, 76*	$C_{12}H_{25}\overset{+}{N}H_3$	*29, 65, 88*
CO_3^{2-}	*66, 76, 77*	$HOCH_2CH_2\overset{+}{N}H_3$	*65*
BO_3^{3-}	*66, 76*	$C_6H_{13}\overset{+}{N}H_3$	*65*
HCO_2^-	*78*	$(HOCH_2CH_2)_3\overset{+}{N}H$	*65*
Tartrate	*50*	$(C_2H_5)_3\overset{+}{N}H$	*10, 60*
Citrate	*79*	*N*-Methylmorpholinium	*25, 89*
$CH_3CO_2^-$	*8, 42, 48, 49, 80–82*	*N*-Methylpiperidinium	*76*
$CF_3CO_2^-$	*60, 83–86*	Piperazinium	*76*
$C_2F_5CO_2^-$	*86*	TEMED	*76*
$C_4F_7CO_2^-$	*76, 86*	Triammonium propane	*76*
$C_6F_{11}CO_2^-$	*86*	$(CH_3)_4\overset{+}{N}$	*65*
$C_4H_9SO_3^-$	*50*	$(C_2H_5)_4\overset{+}{N}$	*65*
$C_5H_{11}SO_3^-$	*66*	$(C_3H_7)_4\overset{+}{N}$	*65*
$C_6H_{13}SO_3^-$	*29, 59, 63, 64*	$(C_4H_9)_4\overset{+}{N}$	*65*
$C_7H_{15}SO_3^-$	*10, 66*	Cationic metal chelates	*90*
$C_8H_{17}SO_3^-$	*10, 66*	Group I and II inorganic cations	*63, 90*
$C_{12}H_{25}SO_3^-$	*59*	Pyridinium	*91–93*
$C_{12}H_{25}SO_4^-$	*29, 59, 63–65*		
Camphor-10 sulfonate	*29, 30*		
p-Toluene sulfonate	*76*		
Naphthalene sulfonate	*76*		
Neutral Polar-Phase Modifying Reagents			
$C_{11}H_{23}CH_2OH$	*10*	$(CH_3)_3CCH_2OH$	*10, 40*
$C_7H_{15}CH_2OH$	*40, 76*	$CH_3CH_2(CH_2)_2COH$	*10, 40*
$C_{10}H_{21}CH_2CN$	*76*		
$C_6H_5CH_2OH$	*40, 76*		

[a] This selection of ionic and neutral species is generally compatible with UV detection below 235 nm, or in the case of those reagents with an aromatic nucleus, with fluorometric or UV (above 254 nm) detection. (The concentration of these components in the mobile phase is usually <5 m*M*.)

[b] For a more complete compendium of applications, conditions, etc., see Ref. *11*.

insulin using a dodecylammonium phosphate system shown in Fig. 9. The same buffer system can be used to resolve bovine insulin from desamido insulin or alternatively this can be achieved with tetramethylammonium phosphate (*17*). Rivier has applied the triethylammonium phosphate system in several studies (*60, 94*) related to hypothalamic releasing factors, whereas other groups have used related acidic amine phosphate buffers (*25, 76, 89*) for the analysis and purification of hormonal peptides.

The most commonly used anions have been dihydrogen phosphate, acetate, and trifluoroacetate. Because of their volatility, the weak organic acids such as acetic or trifluoroacetic acid, usually as the free acid, or the ammonium or pyridinium salts are useful for preparative separations. However, these anions result (*17, 71, 83, 95*) in generally poorer peak shapes compared to phosphate systems and because of their high intrinsic absorbance at 210 nm at levels $>0.1\%$ v/v, they are not suited for very high-sensitivity detection of peptides at their isosbestic point. Deletion of

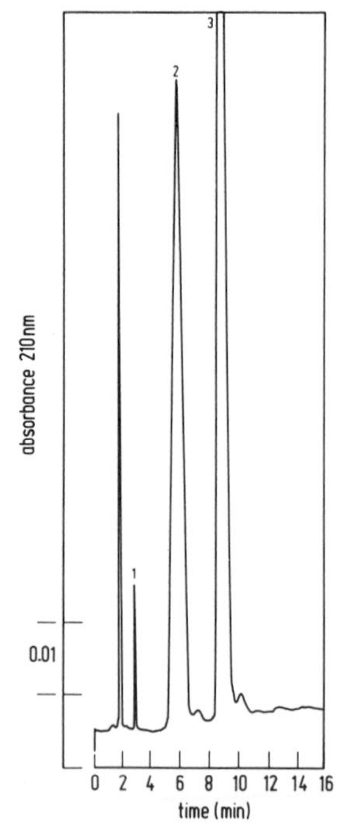

FIG. 9. Separation of porcine despenta-insulin (des-B_{26-31}-insulin) (2) from insulin (1) on a μ-Bondapak alkyl phenyl column using a 60-min linear gradient of 25–50% acetonitrile containing 5 m*M* dodecylammonium phosphate, pH 4.0; flow rate: 2 ml/min. Reproduced from Hearn *et al.* (*70*).

buffer salts with the acetic acid-based eluents results usually in broader peaks and marked loss of resolution (*42, 71*). The use of high-molarity acetic acid–pyridine or formic acid–pyridine mixtures, e.g., 0.5–1.0 *M* mixtures, has been advocated (*57, 91, 92, 96*), but these systems require postcolumn fluorometric derivatization and detection. The retention of peptides eluted from alkyl silicas in the presence of perfluoroalkanoic acids such as trifluoroacetic acid, pentafluoropropanoic acid, and heptafluorobutyric acid show concentration dependencies in accord with predictions based on pairing-ion considerations. Interestingly, these ionic modifiers permit improved recovery for larger, relatively nonpolar peptides which elute in the organic solvent image $0.3 < \psi < 0.6$. Consequently, the perfluoroalkanoic acids have proved useful for the isolation and analysis of larger biological peptides and CNBr peptides (see, for example, Refs. *9a, 86, 97–102*) and in the purification of synthetic peptides on the preparative (i.e., multigram) scale (*103, 104*). It has been the experience of this laboratory that the removal of a nonvolatile buffer component from a recovered peptide sample is usually straightforward in circumstances where it may interfere, e.g., in a bioassay. The usual procedure is to take advantage of the enhanced selectivity due to the most suited counterion and subsequently remove it by rechromatography on a second reversed-phase column with a volatile mobile phase.

A variety of alkylsulfonates and alkyl sulfates have been used as hydrophobic pairing anions in reversed-phase separations of peptides (see Table II). The more hydrophobic of these anions may generate dynamic liquid–liquid cation exchangers. At low pH, retention of peptides is increased in the presence of these reagents with basic peptides exhibiting the largest increases. As a consequence, major changes in selectivity can be achieved. In view of the detergent-like properties of many of the hydrophobic anionic pairing ions, disaggregation of peptide–peptide noncovalent complexes occurs. A dramatic improvement in peak shape and recovery of larger peptides is often seen (*29, 50, 60, 63, 64*) with these anions, a situation which can also be found with hydrophobic alkylammonium cations. The chromatograms shown (*29*) in Fig. 10 illustrate the effect of pairing ions on the separation of the peptides Ala-Tyr, Pro-Tyr, and Leu-Tyr. The separation of synthetic adrenocorticotropin analogs with a 1-butane–sulfonate buffer system has been reported (*50*). These conditions permit the resolution of α_p-ACTH and α_H-ACTH (the only difference being Leu_{31} in α_p-ACTH and Ser_{31} in α_H-ACTH). Assessment of peptide homogeneity can be readily made when two or more appropriate pairing-ion elution systems are employed sequentially. Such approaches have been advocated for "two-dimensional" peptide mapping as well as for isolation of biological peptides, and in particular for the identification

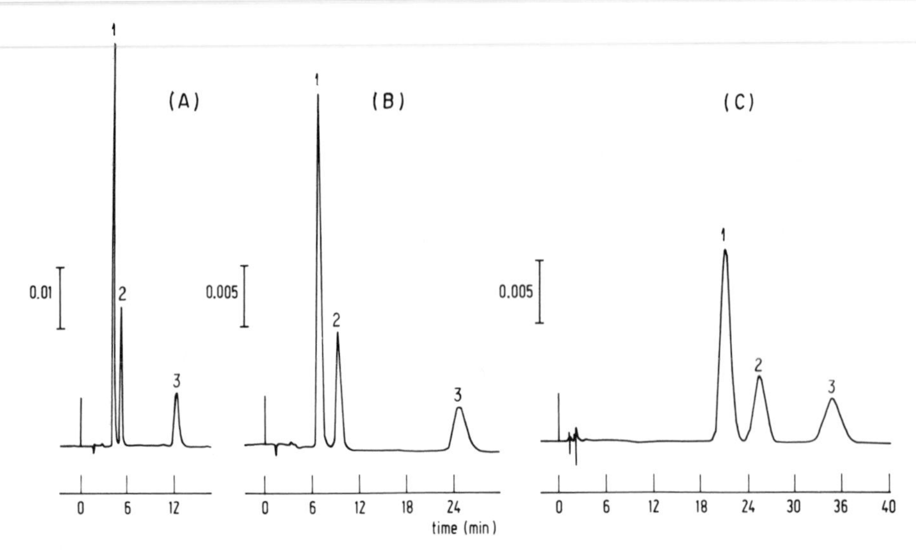

FIG. 10. Chromatograms illustrating the effect of pairing ions on the separation of the dipeptides Ala-Tyr (1), Pro-Tyr (2), and Leu-Tyr (3) by reversed-phase chromatography. Column: μ-Bondapak C_{18}; flow rate: 2 ml/min; temperature: 20°C; eluents: A, 5% methanol–95% water–50 m*M* KH_2PO_4–5 m*M* H_3PO_4, pH 3.0; B, same as A plus 5 m*M* *n*-hexyl sulfonate; C, 50% methanol–50% water–50 m*M* KH_2PO_4–5 m*M* H_3PO_4–5 m*M* dodecyl sulfate, pH 3.0. Reprinted with permission from Hearn *et al.* (*29*). Copyright by Elsevier Scientific Publishing Co., Amsterdam.

of peptide fragments from biosynthetic studies on radioisotopically labeled precursor proteins (*11, 29, 30*).

A logical extension of buffer ion interactive phenomena in the reversed-phase HPLC of peptides is ligand-exchange chromatography. The potential of ligand-exchange chromatography for the separation of amphoteric molecules, such as amino acids, is well documented. To date, application of ligand-exchange methods to peptide resolution has attracted much less attention due in no small measure to the capabilities which conventional reversed-phase HPLC approaches presently offer. Because outer-sphere metal complexation with divalent or trivalent cations such as Cu^{2+}, Zn^{2+}, or Co^{3+} offers future scope for control over selectivity, their application to peptide analysis warrants further investigation.

H. Effect of Temperature and Flow Rate

At this stage, both temperature variation and solvent flow programming as means of controlling separations of peptides on reversed-phase columns have been poorly investigated. If we recall Eq. (12) then the dependence of k' on T can be expressed as

$$\log k_i' = -\frac{\Delta H^\circ_{\mathrm{assoc},i}}{RT} + \frac{\Delta S^\circ_{\mathrm{assoc},i}}{R} + \log \Phi \tag{21}$$

where $\Delta H^\circ_{\mathrm{assoc},i}$ and $\Delta S^\circ_{\mathrm{assoc},i}$ are the standard enthalpy and entropy changes for the transfer of the solute S_i to the stationary phase, respectively. Although hydrophobic interactions have been postulated to increase with increasing temperature, in fact the capacity factors of peptides in reversed-phase separations are generally reduced on raising the temperature. Minor selectivity changes have been noted (*17, 60*) for peptides with temperature variation. This dependency may fall into the class of so-called "entropy-dominated" temperature effects which have been suggested to correlate with differences in molecular shape of solute molecules (*105*). Higher temperatures will favor more rapid mass transfer of the peptidic solutes and this will lead to improved efficiencies, although the possibility of solute and stationary-phase degradation at higher temperatures must be taken into account. Often a useful increase, e.g., 2-fold, in column efficiency can be obtained by raising the temperature from ambient to 40–45°C. However, Rivier has concluded (*60*) on the basis of elution studies with luteinising releasing factor (LRF) that even when the elution volume of the peptide is made constant at two different temperatures, selectivity at the higher temperature is poorer than at the lower one.

It has been well recognized in HPLC that column efficiency is dependent on linear flow velocity of the mobile phase and this dependency is usually visualized in the form of Knox plots. Improved efficiencies (and reduced retention times) are possible under elution conditions involving flow programming, but this technique has been little used for peptide separations since it is better suited to solutes exhibiting a narrow range of polarities. Under isocratic conditions an increase in flow rate results in a corresponding decrease in the capacity factors of peptides, although the elution volumes remain essentially constant. Similarly, under conditions of gradient elution, flow rate changes make only small differences to the gradient elution time provided the rate of change of the organic solvent modifier is low, i.e., $<1\%$/min. For analytical-type columns (ca. 25 cm $\times$ 4 mm) packed with alkyl silicas for which $d_p = 5$ or 10 μm, flow rates of 0.5–2.0 ml/min are generally used, although for preparative separations with the radially compressible columns flow rates of ca. 100 ml/min or higher can be used (*55*).

In a comparative study on the influence of linear flow velocity and organic modifier composition, it was found (*17b*) that the separation efficiencies for small peptides, as expected on theoretical grounds, decreased with increasing solvent viscosity at fixed flow rates. However, with larger peptides/polypeptides, divergences have been noted where some polypeptides exhibit poorer separation efficiencies in eluents of lower viscosity. This trend was most notable between methanol and propan-1-ol and may reflect solubility parameter and solvation differences (*17b,c, 30*).

I. Separation of Diastereoisomeric Peptides

Racemization of chiral centers during the isolation or synthesis of a particular peptide will lead to diastereoisomeric mixtures. In addition, in peptide synthesis it is often more efficient to use DL-amino acids, particularly if they are specifically labeled with ^{13}C, ^{14}C, ^{3}H, ^{35}S, ^{125}I, etc. Rapid and reproducible chromatographic methods for separation of diastereoisomers at the analytical and preparative level are thus desirable. Several groups of investigators have used polar adsorption HPLC on silica for the adsorption of fully protected diastereoisomeric peptides (see Section IV), whereas with unprotected peptides the classical procedures of countercurrent distribution, partition chromatography, and ion-exchange chromatography have been used. Because diastereoisomers have different hydrophobic surface areas, they are resolvable by reversed-phase HPLC. In fact, retention will be dependent on the orientation of the side-chain groups on the asymmetric α-carbon atoms with regard to the peptide backbone. This can most easily be visualized by considering the stereochemical arrangements of the isomeric pair L-X–L-Y and L-X–D-Y where X and Y are the same or different amino acids. As has been shown from a number of physicochemical studies, including nuclear magnetic resonance (*106*), ionization constants (*107*), and molecular size and dipole moment measurements (*108*), the amino acid side chains adopt a transoid arrangement in the L–L (and its D–D enantiomer) isomer and a cisoid arrangement in the L–D (and its D–L enantiomer) isomer. Thus for the diastereoisomers L-Phe–L-Phe and L-Phe–D-Phe, the cisoid arrangement of the nonpolar phenylalaninyl moieties in the L–D isomer will produce a larger overall surface area and increase the retention compared to the L–L isomer, where the side chains are on opposite sides of the peptide bond, i.e., k' for L-Phe–L-Phe is smaller than k' for L-Phe–D-Phe, and a similar circumstance will hold for the diastereoisomeric dipeptides of the other nonpolar amino acids. If the amino acid side chain contains a polar ionized moiety, e.g., as in the diastereoisomeric pair L-Leu–L-Arg and L-Leu–D-Arg at low pH, an elution order reversal is expected with the L–D isomer now eluting more rapidly than the L–L isomer. Experimental confirmation of these conclusions can be found in a number of studies (*7, 25, 46, 48, 49, 80, 109–111*).

Similar stereochemical arguments can be extended to larger peptides. For example, the elution order of diastereoisomeric nonpolar peptides of the type L-X–L-Y–L-Z, L-X–L-Y–D-Z, L-X–D-Y–L-Z can be predicted on conformational grounds to be LLL < LLD < LDL and this has been observed experimentally, e.g., enkephalin analogs (*112*). Furthermore, the introduction of the glycinyl residue into an endo position between two chiral amino acids decreases both the k' and α values for the diastereoisomers (*46*).

These principles have been applied in a number of studies on the reversed-phase HPLC separation of diastereoisomeric peptides, including simple hydrophobic dipeptides (*46, 109*), hypothalamic releasing factors (*28, 113, 114*), oxytocin analogs (*80, 111*), and somatostatin analogs (*7, 25*). Illustrative of these separations is the resolution of bombesin from [D-Met_{14}]-bombesin (*49*) shown in Fig. 11. Since peptide diastereoisomers of more than 10 amino acid residues are generally very difficult to separate by chemical techniques, these reversed-phase HPLC methods represent a major improvement in separative capability. It is well known from other studies that the replacement of an amino acid residue by its enantiomer has an appreciable effect on the molecular conformation of a peptide even when the epimeric position is terminal. Depending on the kinetics, other stereochemical conformations, e.g., rotational or ring inversion conformers in solution, may also have an effect on retention and these will be manifested as selectivity changes or multiple peaks dependent on the nature of the nonpolar stationary phase and composition of the mobile phase. At this stage, these aspects have been poorly explored, but it is likely that HPLC with nonpolar stationary phases could become a useful

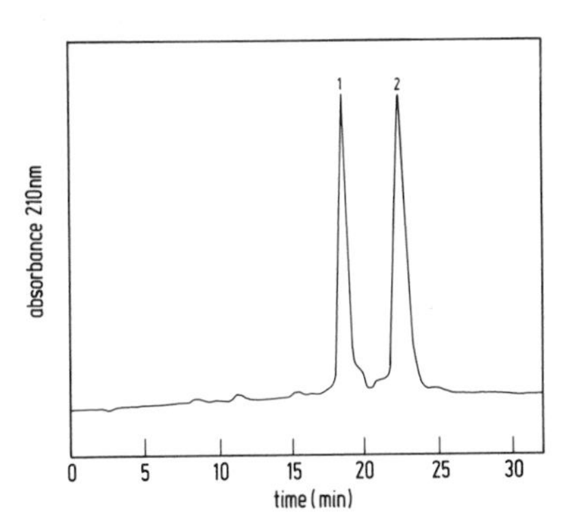

FIG. 11. Separation of bombesin (1) from [D-Met_{14}]-bombesin (2) by HPLC. Conditions: μ-Bondapak C_{18} column (120 × 0.7 cm); flow rate 6 ml/min; isocratic elution using 30% acetonitrile–10 mM ammonium acetate, pH 4.2. Reprinted with permission from Rivier and Brown (*49*). Copyright by American Chemical Society.

tool to probe the thermodynamic and kinetic equilibria associated with conformational changes as a function of structure, pH, organic solvent concentration of the mobile phase, etc.

Similarly, reversed-phase HPLC can be used as an alternative to the racemization test for amino acids as developed by Manning and Moore (*115*). Rivier and Burgus (*109*) have suggested the use of L-phenylalanine, coupled via the *N*-carboxyanhydride method to a hydrolysate, to monitor racemization during synthesis, although other hydrophobic L-amino acids should also prove equally effective. The use of *tert*-butyloxycarbonyl-L-amino acid-*N*-hydroxysuccinimide esters in the separation of enantiomeric amino acids and diastereoisomeric peptides has been described (*110*). Ultimately, these methods may not prove as versatile as the use of chiral stationary phases made by stereoselective control of the bonding process or, alternatively, with surface-active reagents similar to the D-4-dodecyldiethylenetriamine-Zn(II) complex recently introduced by Karger *et al.* (*90*) for the outer-sphere ligand-exchange chromatography of enantiomeric dansyl amino acids.

J. Analysis and Assessment of Homogeneity of Natural and Synthetic Peptides on Reversed Phases

In the previous sections, the influences of a number of chromatographic and structural parameters on the retention of unprotected peptides to reversed phases was examined and their effect rationalized on the basis of solvophobic considerations. Similar considerations can be applied to protected, or partly protected, peptides as derived from solution or solid-phase synthesis or by chemical derivatization of a naturally occurring substance. In these cases, it may not be possible to take advantage of protic or other secondary equilibria due to the removal of ionogenic groups. Furthermore, most blocking groups used to protect potentially reactive functional sites in a peptide are in their own right usually nonpolar and their presence will further decrease the polarity of the peptide compared to its fully unprotected form.

Fully protected peptides can be successfully chromatographed on both reversed phases and silica gels (see Section IV). For the reversed-phase separation of fully protected peptides, mobile phases containing high volume fractions of organic solvents may be required, e.g., ψ values in the range 50–70%. Elution is generally carried out without pH or buffer control. Knox and Szokan have used (*116*) an ODS-TMS silica to analyze a preparation of protected Leu-enkephalin, Z-Tyr-Gly-Gly-Phe-Leu-OMe, and its base hydrolysis products, including the free acid, using a methanol–water–acetonitrile (3 : 2 : 4) system. The free acid, being more polar, elutes before the methyl ester. Nachtmann has reported the analysis of the protected peptides formed during the synthesis of oxytocin using isocratic elution and an octyl silica support (*117*). Similar reversed-phase approaches have been used for the separation of protected or partially protected peptides related to the enkephalins (*8*), acyl carrier protein (*63*), and type C viral antigens (*118*). Because the functional group contributions of both the amino acid side chains and the protecting groups will dominate retention with fully protected peptides, it is possible to design a synthesis around a particular protecting group so that subsequent analysis and purification of the peptidic product is simplified.

With careful choice of elution conditions and UV detection at 210 nm or fluorometric detection, the detection limit for impurities in a peptide sample is frequently better than 1% w/w. Because of the excellent control over selectivity which is possible with reversed-phase HPLC, it is usually possible to obtain baseline resolution of peptides of very similar sequence e.g., peptides differing by only one residue, desamido isomers, etc. Purity criteria can be based (*8, 64, 65*) on the use of several different elution conditions analogous to those derived from classical thin-layer chromatography or paper chromatographic methods. With the reversed-phase HPLC methods, solution and storage stability of both protected and unprotected peptides can be easily monitored and differences due to the source or the method of manufacture ascertained. With the increasing usage of synthetic peptides as therapeutic substances, particularly in the area of endocrine dysfunction, this knowledge on the level of purity becomes of utmost importance. The use of reversed-phase HPLC in the analysis of several pharmaceutically important peptides has recently been reviewed (*9*). Notable examples of the application of reversed-phase HPLC in this area include oxytocin, lypressin, and ornipressin (*47, 53, 80, 95, 111, 119, 120*); neuroendocrine-releasing factors and their metabolites (*28, 50, 60, 93, 114, 121, 122*); the angiotensin family (*8, 33*), insulins (*24, 50, 63, 95, 123, 124*); glucagons (*28, 63, 67, 95, 125*); adrenocorticotropin-related peptides (*17, 50, 126*), calcitonin-related peptides (*122*), and growth-hormone-related peptides (*9b, 52f*). Furthermore, similar methods can be used for the analysis and purification of radioisotopically labeled peptides for use in radioimmunoassays and radioligand binding assays.

Besides these specific examples related to peptides of therapeutic importance, similar thematic variations can, in general, be used to (1) assess the purity of natural and synthetic peptides isolated by alternative procedures; (2) analyze the crude reaction product mixtures from solution- or solid-phase synthesis, thus allowing deletion or partially deprotected products to be rapidly recognized and the progress of a synthesis to be followed; (3) confirm the identity of a peptide by co-chromatography with a purified standard under several elution conditions, and (4) allow the preparative recovery of the desired peptides. Illustrative of these various facets of the use of reversed-phase HPLC for peptide separation are the studies encompassing peptide fragments of β-thyrotropin (TSH) and antamanide (*8*), angiotensin fragments (*8, 27, 33, 63*), bombesin and its analogs (*7*), intestinal peptides (*113*), fibrinogenic peptides (*127*), enkephalinic peptides (*8, 126, 128–131*), oxytocin-related peptides of pituitary origin (*132, 132a, 133*), insulin-related peptides (*24, 63, 67, 72, 88, 134*), ACTH-related peptides (*63*), peptidic antibiotics (*33, 73, 135*), α- and β-melanotropin (MSH) peptides (*136–138*), renin inhibitor (*139*), substance P (*28, 140*), somatostatins (*7, 84, 94, 109, 114, 141, 142*), endorphin-related peptides (*28, 60, 83, 89, 109, 126, 143–147*), and cholecystokinin peptides (*148*).

Provided consideration is given to the nature of the weak dispersion forces involved in the sorption phenomenon, efficient and reproducible chromatography can be carried out with polypeptides and with proteins of MW up to ca. 680,000, using reversed phases of suitable particle and pore diameters and ligand coating. With many polypeptides the range of ionic strengths over which the "salting in–salting out" effects favor desorption from an alkyl silica is narrow and inappropriate buffer concentrations can lead to extensive retention or peak broadening. Similarly, band spreading will arise due to poor mass transfer with supports of small pore sizes. Furthermore, many polypeptides can only be efficiently chromatographed on reversed phases over a narrow range of organic solvent concentrations. Although advantage is taken of this effect in gradient elution, some care in the selection of gradient shape and duration is necessary to ensure that the narrow desorption range of elutropic strength for the mobile phase is not exceeded too rapidly. Above this range, loss of selectivity and even enhanced resorption of the polypeptide may occur onto the stationary phase. Several parameters can be chosen to limit the strong interaction of polypeptides with alkyl silicas. These include the use of reversed phases of large poor diameter and surfaces bonded with a less hydrophobic ligand; reduced flow rate; decreased rate of change of an organic solvent modifier such as acetonitrile or 1-propanol; and the use of ternary-solvent systems. The major concern with these reversed-phase separations of biologically active polypeptides, in addition to that of the mass recovery yield, is the possibility of mobile- or stationary-phase-induced conformational changes leading to loss of biological or immunological activity. The extent to which this occurs will partly depend on the lability of the polypeptide, which will be different in each case. However, recent experiences have shown that these potential limitations can be largely overcome. Application of these chromatographic strategies to large polypeptides and proteins have been described elsewhere (*11–13*), but it is useful to note with the following selected examples that the separation of macroglobulins on reversed phases encompasses a very large range of molecular weight. Typical of the analytical and micro-preparative separations now feasible with larger polypeptides and proteins are the studies on insulins isolated from different pancreases (*24, 88, 123*) or from the galactosidase–insulin chain hybrid proteins (*149*), relaxins (*77*), human leukocyte interferon (*81*), pituitary protein hormones (*10, 39a, 82*), the neurophysins (*133*), the bovine pancreatic trypsin inhibitor aprotinin (*150*), ferritin (*134*), collagen chains (*43*), and even thyroglobulin (*44*).

K. Preparative Separations of Peptides on Reversed Phases

Columns and reversed-phase HPLC packing materials suitable for sample loadings ranging from 1 ng to 10 gm or higher are now available either commercially or laboratory made. Separations with so-called analytical systems using standard stainless steel columns (e.g., 25 cm × 3.9 mm) packed with alkyl silicas, for which d_p = 5 or 10 μm, can be carried out with sample loadings of up to, and in some cases higher than, 10 mg depending on the selectivity factors obtained, i.e., the larger the α differences the larger the sample load per injection. In circumstances where a particular peptide is available in low abundance, highly efficient preparative separations can be carried out on analytical columns. The peptides can be detected at high sensitivity spectrophotometrically, e.g., at 210 nm at 0.005 AUFS, or by bioassay, radioimmunoassay, or radioligand assay of the recovered fractions. In many cases, the peptide samples, as directly obtained or following lyophilization, can be used in these assays. The presence of residual amounts of simple buffers, such as KH_2PO_4, NH_4OAc, and NH_4COOCF_3, does not appear to adversely effect radioimmunoassays or radioligand assays down to low detection levels. For example, less than 20 pg of immunoreactive β-endorphin can be detected (*151*) per milliliter of thyroid extracts following reversed-phase HPLC. Because high concentrations of buffer salts interfere in some bioassays due to perturbation of the osmolarity, a desalting step may be required. This can often be readily achieved by rechromatography on another reversed-phase column using volatile buffers as the eluent.

Examples of the use of analytical-scale column systems for the small-scale (i.e., 1 μg to 1 mg) preparative separation of peptides include the extraordinarily potent opoid peptide, dynorphin (*145*); insulin A-, B-, and C-chain peptides (*24, 72*); β-chain peptides of the pituitary glycoprotein hormones (*8*); endorphins (*28, 60, 126, 137*), adrenocorticotropic peptides in plasma, pituitary, and other endocrine glands or secreted from tumor cells *in vitro* (*84, 125, 127, 138, 142, 151*); hypothalamic releasing factors

(*114, 121*); rat β-lipotropin (*82*); human leukocyte interferon (*81*); camel pro-opiocortin, the common precursor of the opoid peptides and corticotropin (*92*); relaxin β-chain peptides (*77*); enkephalin precursors (*129, 131*); somatostatins in rat plasma and pigeon pancreas (*84, 94, 114, 142*); α- and β-melanotropin (*137, 138*); hemoglobin α-, β-, and γ-chains (*152*); calcitonins (*151*); parathyroid hormone fragments (*151*); vasopressin precursors (*28a*); substance P (*140*); secretin peptides (*153–155*); bradykinin activator peptides (*156*); phosphopeptides from oncoproteins and corticotropin-related precursors (*157, 158*); amunin (*159*); growth factors (*9a, 160–163*); neurophysins (*164*); and a variety of enzymatic or CNBr-derived peptides generated during fragmentation studies (see p. 134).

Intermediate-scale (or semi-) preparative separations of peptides with sample loadings usually in the range 500 μg–10 mg per injection have been carried out with the columns of increased internal diameters or alternately with several analytical-scale columns joined in tandem. In this laboratory either larger stainless steel columns (60 cm × 7 mm, 90 cm × 7 mm), three to four analytical columns in tandem, or the small, radial compressible Teflon columns (10 cm × 10 mm) have been routinely used, in some cases with sample loadings up to 50 mg peptide per injection. Because of the higher, and at times limiting, backpressures obtained with semi-preparative columns packed with 5- or 10-μm particles, the larger fully porous packings with d_p 20–50 μm are useful alternatives with these bigger columns. Their use does lead to some loss of resolution but with these packing materials the columns can be dry-packed by the tap-fill method (*165*). Shown in Fig. 12 is the elution profile for the purification of the crude Leu-enkephalin amide O-benzyl ester on a phenyl Porasil B column eluted with acetonitrile–water–orthophosphoric acid (30 : 70 : 0.1% v/v). Other examples of the semipreparative purification of peptides and polypeptides include the diastereoisomeric [D-His_2]-TRF and TRF (*121*); ribonuclease (*13*); insulins from animal pancreases (*88*); the hexadecapeptide α-endorphin from pituitary extracts (*28*); pigeon pancreatic somatostatin (*94*); the protected (1–21)-peptides of baboon and cat RNA tumor viral P_{30} antigens (*118*); fully protected [1-5]ACP (*63*); growth hormone and growth factor-related peptides (*9a, 161, 162*); and amunin (*159*).

Although a vast array of peptides can now be synthesized for use in biological investigations, a major difficulty with these studies has been the inability to rapidly isolate high-purity underivatized peptides in gram quantities. Similar production limitations with the more abundant naturally occurring peptides are also obvious. Several studies have described the use of large radially compressible Teflon cartridges (30 × 5.5 cm), containing 10- to 20-μm octadecyl silica particles for the separation of peptides with sample loadings up to 10 gm per injection. With these

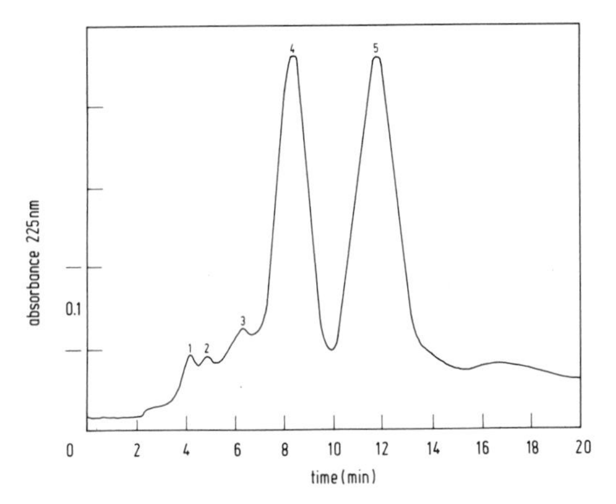

FIG. 12. Semipreparative separation of crude Leu-enkephalin amide *O*-benzyl ether (5) on a phenyl Porasil B column (60 × 0.7 cm) using a 30% acetonitrile–70% water–0.1% orthophosphoric acid isocratic mobile phase at a flow rate of 2 ml/min. The loading was 4 mg of peptide. Reprinted with permission from Hearn *et al.* (*8*). Copyright by Marcel Dekker, Inc., New York.

systems, flow rates are high (e.g., 100 ml/min) and efficient recycle chromatography is feasible. Typical of these separations is the elution profile for the preparative purification (*85*) of 1 gm of crude L-Leu-Gly-Gly-Gly shown in Fig. 13. In this case only 21 of a methanol–water–trifluoroacetic acid (5 : 95 : 0.05% v/v) mobile phase was required and elution, including recycle, was completed in 27 min. Similar condi-

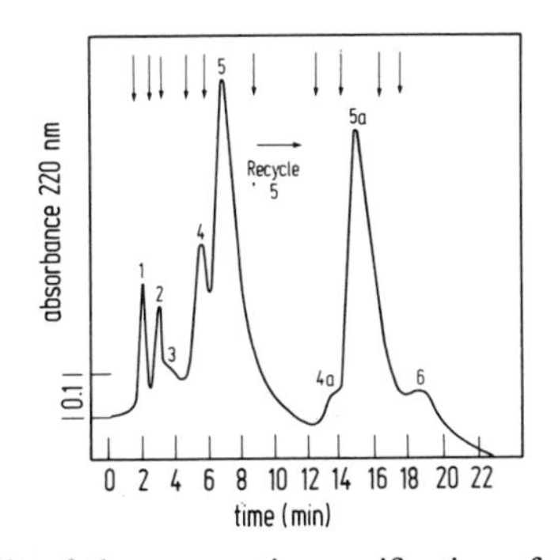

FIG. 13. Elution profile of the preparative purification of 1 g of L-Leu-Gly-Gly-Gly. Chromatographic conditions: column, PrepPak-500/C_{18} cartridge; mobile phase, 95% water–5% methanol–0.05% trifluoroacetic acid, pH 2.3; flow rate, 100 ml/min. The arrows indicate where fractions were pooled; fraction 5 was recycled once and the desired fraction 5a was collected. Reprinted with permission from Bishop *et al.* (*103*). Copyright by Elsevier Scientific Publishing Co., Amsterdam.

tions have been used for the large-scale purification of somatostatin (*114*) and enkephalin-related peptides (*103*).

Because of the favorable sorptive properties of the reversed-phase supports, batch adsorption and desorption can be a very effective way to desalt a chromatographed sample or to partially fractionate a peptide mixture during a purification procedure. For example, 1–2 gm of an octadecyl silica packed into a silanized glass or plastic pipette can be used for the batch fractionation of small amounts of a crude peptide extract from tissues, such as the pancreas or pituitary, or from a synthetic experiment. A number of commercial products, such as the Waters Sep-Pak, have found use in this manner (*10*) as a purification or sample preparation aid. Protocols for batch extraction procedures on alkyl silicas have been discussed (*17a,b*) and applied to neuropeptides (*10, 158, 166*) and other hormonal peptides (*88, 162, 167, 168*). With these methods recoveries of peptides present in a "tissue" extract are generally higher than those found with classical fractionation techniques due in part to the fact that proteolytic degradation is minimized.

L. Structural Mapping of Polypeptides and Proteins on Reversed Phases

One of the most powerful demonstrations of the capability of reversed-phase HPLC for the separation of peptides is found in its use for the structural mapping of polypeptides and proteins. Compared to the conventional techniques (open-column gel or ion-exchange chromatography, two-dimensional paper or thin-layer chromatography or electrophoresis in one direction followed by chromatography in a perpendicular direction) that have been used in "peptide mapping" studies, reversed-phase HPLC offers superior peak resolution, shorter elution or analysis times, and improved recoveries. With careful selection of the chromatographic conditions in reversed-phase HPLC, optimal selectivity factors can be achieved for peptides generated by enzymatic or chemical fragmentation of polypeptides and proteins. In addition to their direct value in structure determination, these methods permit proteins to be compared, distinctive sequence homologies to be detected, and characteristic protein variants to be identified. The fact that reversed-phase HPLC methods have a resolving power at least one order of magnitude greater than conventional methods has significant ramifications for the separation of peptide mixtures from larger macroglobulins, e.g., for proteins like human thyroglobulin, which has potentially ca. 470 tryptic peptides and ca. 63 expected CNBr peptides (*169*). In these cases, complete resolution by conventional methods is totally impracticable, but is potentially attainable with reversed-phase HPLC methods. Several strategies for the use of reversed-phase HPLC for the separation of peptide fragments from larger naturally occurring molecules have recently been reviewed (*74, 91*). These approaches permit reliable analytical peptide mapping of polypeptides and proteins, the detection of cleavage intermediates, and the time course of the cleavage reaction to be followed. Because of the preparative capability, these methods frequently allow the direct recovery of the eluted peptides in a form suitable for subsequent amino acid composition or sequence determination. Similar approaches can be applied to large synthetic peptides to confirm the accuracy of a synthesis. For reasons made obvious in earlier sections, gradient elution with aquo-organic solvent mixtures of low pH and containing phosphate, acetate, formate, or trifluoroacetate buffers has most frequently been used for both analytical and preparative separations. Elution under two mobile-phase conditions of identical organic solvent composition and pH, but differing in regard to the polarity of a counterionic pairing ion, has been used (*29*) to provide pairs of chromatograms formally analogous to those obtained with peptide maps by two-dimensional electrophoretic techniques.

The advantages of phosphate-mediated elution systems at pH 2–3 have been widely utilized following the initial report (*170*) on the analysis of peptides formed by the tryptic or thermolysin digestion of a number of globular proteins, including acyl carrier protein, the thiol protease actinidin, and ribonuclease S peptide (*33*). With the aquo-acetonitrile gradient systems (0.2–1.0%/min at flow rates of 0.5–2.0 ml/min) containing 0.1% orthophosphoric acid, peptide mapping of 10 n*M*, or smaller, quantities of proteins is possible with high-sensitivity UV detection at 200–210 nm. Dual-wavelength monitoring at 210/254 nm or 210/278 nm provides useful information on the aromatic amino acid content of the eluted peptides. The elution profile for the tryptic digest of human growth hormone shown in Fig. 14 typifies the excellent resolution which can be achieved with reversed-phase supports and acetonitrile–water–0.1% orthophosphoric acid gradient elution systems. Similar chromatographic conditions have subsequently been applied to the separation of the tryptic peptides of hemoglobin variants (*10, 55, 67*), pituitary glycoprotein hormones (*171*), immunoglobulin γ-(*171*) and μ-chains (*172*), bovine and chick intestinal calcium-binding proteins (*75*), the enzyme phosphofructokinase (*67*), human and sheep thyroglobulin-related proteins (*6*), and peptide prohormones (*24*). Detection sensitivity for specific peptide fragments can be substantially increased with *in vivo* or *in vitro* radioactive labeling of polypeptides or proteins. Both ^{3}H- and ^{14}C-labeled peptides coelute with their isotopically stable equivalent, whereas ^{125}I-labeled peptides tend to show longer retentions than their noniodinated equivalent on reversed

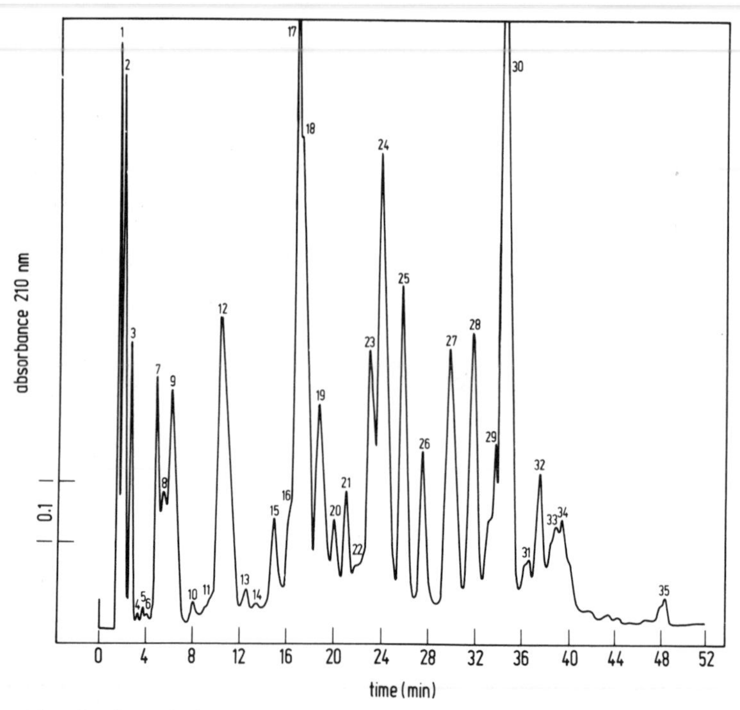

FIG. 14. Gradient elution profile of the tryptic digest peptides of human growth hormone. Chromatographic conditions: column, μ-Bondapak alkyl phenyl (30 × 0.4cm); flow rate, 2.0 ml/min; temperature, 20°; elution conditions, a 60-min linear gradient was generated from water–0.1% H_3PO_4 to 50% acetonitrile–50% water–0.1% H_3PO_4; detection, 210 nm, 0.2 AUFS with electronic subtraction of the gradient background. Reproduced from Hearn *et al.* (*163*).

phases with phosphate-based methanol–water or acetonitrile–water gradient elution systems. Separations based on these conditions include the radiolabeled tryptic peptides of rat caseins (*173*), of influenza-related glycoproteins (*174*), and of *in vivo* labeled guinea pig thyroglobulin (*6*). A similar phosphate approach with an elution gradient of acetone has been used by McMillan *et al.* (*175*) for the separation of the labeled tryptic peptides of the α- and β-polypeptides of the Ia antigens from the I-E subregion. The triethylammonium phosphate system has been used (*60*) for the analysis of the tryptic digest of myelin basic protein.

The use of low concentrations (0.1–5%) of acetic, formic, or trifluoroacetic acid, generally with gradient elution, has been reported for the separation of tryptic or CNBr peptides on alkyl silicas. For example, the CNBr peptides of the C-1 chymotryptic fragment of bacteriorhodopsin have been separated (*78*) on a μ-Bondapak C_{18} column using 5% formic acid and gradient elution from 40–80% ethanol. Other examples of the use of low concentrations of the organic acids include the separation of the tryptic peptides of doubly (^{3}H- and ^{14}C-) labeled α- and β-subunits of murine I-A alloantigens (*176*) (Fig. 15), of the tryptic peptides of lysozyme, and of the CNBr peptides of the avian virus P_{27} structural protein and the influenza virus M protein (*172*).

More frequently, the alkanoic acids have been used as ammonium alkylammonium or pyridinium salts in reversed-phase HPLC peptide mapping studies. Because of their UV opacity postcolumn derivatization with fluorochromic reagents and fluorescence monitoring of the enzymatic or CNBr digest is required with pyridine acetate, formate, etc. Generally, pyridine-based buffers have been used at high molarities with propanol gradient elution. For example, the identity of rat β-endorphin with camel β-endorphin was confirmed (*91*) via their common tryptic peptides separated with a 1 *M* pyridine–0.5 *M* acetic acid buffer and a linear gradient of 0–20% 1-propanol on a LiChrosorb RP18 column (Fig. 16). Similar methods have been used to resolve the tryptic peptides of the following proteins: ovalbumin (*91*), the putative enkephalin precursors found in bovine adrenal medulla (*5*), myoglobin (*96*), pro-opiocortin (*92*), the CNBr peptides of the MOPC-315 mouse immunoglobulin heavy chain (*177*), and the α-chain goldfish hemoglobin (*178*).

The reversed-phase HPLC separation of the tryptic peptides of hemoglobin variants using ammonium acetate buffers, generally 10 m*M* at pH 6.0–6.7, with methanol, acetonitrile, or propanol gradient elution has been reported by several groups. Despite the poorer resolution (often broader or overlapping peaks) obtained with acetate compared to orthophosphate

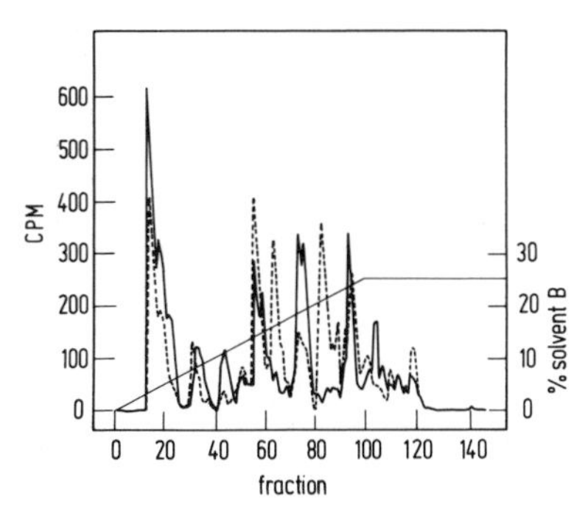

FIG. 15. Separation of the tryptic digest peptides of the murine Ia alloantigen A^k (——) and A^r (-----) β-polypeptides on a μ-Bondapak C_{18} column using a 2% acetic acid primary eluent and a linear 60-min gradient generated from 0 to 25% acetonitrile at a flow rate of 0.7 ml/min. Reprinted with permission from Cook *et al.* (*152*). Copyright by Williams and Wilkins Co., Baltimore, Maryland.

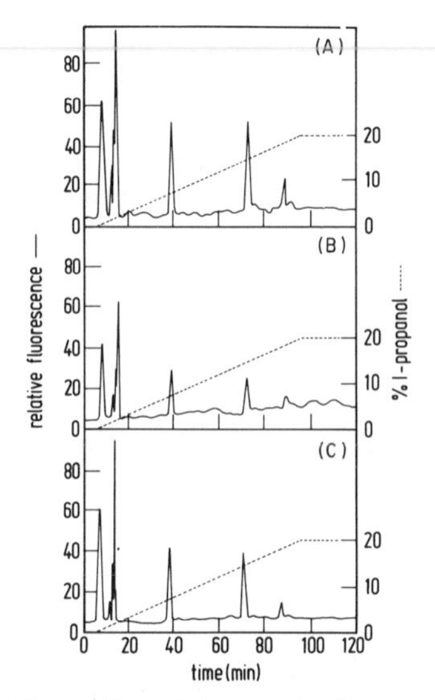

FIG. 16. Chromatographic profiles of the tryptic digest peptides from rat and camel β-endorphin. Chromatographic conditions: column, 10-μm LiChrosorb RP18 (25 $\times$ 0.32 cm); eluent, 1 *M* pyridine–0.5 *M* acetic acid primary buffer with a 100-min linear gradient of *n*-propanol from 0 to 20%; flow rate, 14 ml/h. (A) Mixture of trypsin digested rat (150 pmol) and camel (250 pmol) β-endorphin; (B) rat (150 pmol) tryptic peptides; (C) camel (250 pmol) tryptic peptides. Reprinted with permission from Rubinstein *et al.* (*77*). Copyright by Academic Press, Inc., New York.

systems, the identification of the variant hemoglobin peptide is still usually straightforward since the composition of all the α- and β-chain tryptic peptides is well documented. These methods have been used to separate tryptic peptides of an apohemoglobin variant subsequently identified as HβE (*87*), the tryptic and CNBr peptides of HβF, A, A_2, and variants (*179–182*). Similar conditions have been described for the separation of the tryptic peptides of β-endorphin (*113*), the CNBr cleavage products from the β-galactosidase-insulin A- or B-chain hybrid proteins (*149*), and the CNBr peptides of vasoactive intestinal polypeptide and human pancreatic polypeptide (*113*).

Alkylammonium acetate buffer systems have also attracted limited attention for the separation of tryptic digests of proteins on octadecyl silica (*60*). The ammonium formate or trifluoroacetate buffers have also been used in the reversed-phase HPLC separation of the tryptic, chymotryptic, and carboxypeptidase A digests (*136*) of α-melanotropin and its *N,O*-diacetylserine analog as well as for enzymatic digests of the adrenocorticotropic family (*137*). Ammonium bicarbonate eluent systems have proved useful in mapping studies due to the eluent's volatility and different selectivity patterns compared to low-pH acid systems. However, peptide recoveries tend to be lower with bicarbonate systems than those generally found with phosphate or trifluoroacetate. Also, some alkyl silica supports do not exhibit satisfactory long-term stability. One percent ammonium bicarbonate with acetonitrile (or propanol) gradients have been used successfully with the radially compressible, flexible walled plastic columns packed with octyl (or octadecyl) silicas (Waters RCM supports). These chromatographic supports have been employed (*183, 184*) for the separation of hydrophobic tryptic peptides, e.g., tryptic peptides from apolipoproteins and immunoglobulins, at the semipreparative level (typically 1–10 mg) on standard commercial (10 × 0.8 cm) columns.

In addition to the examples discussed above, these various reversed-phase HPLC mapping procedures have subsequently found numerous other advocates. Selected recent achievements include application to human hemoglobin variants (*9a, 182, 185*), the α- and β-chains of rat hemoglobin (*186*), polypeptide hormones (*9a, 99, 163*), porcine C5a anaphylatoxin (*187*), *Aplysia* neuroactive polypeptide (*188*), ACTH/β-lipotropin precursors (*97, 99, 189, 190*), oncoproteins (*157*), chick liver dihydrofolate reductase (*191*), limulin (*192*), ATP-citratelyase (*193*), spore-specific proteins (*194*), cAMP-dependent protein kinases (*195, 196*), interferons (*197*), complement components (*9a, 198*), α_2-macroglobulin (*198a*), lectins (*199*), phycobiliproteins (*200*), bovine mitochondrial-F, ATPase (*201*), collagens, tubulins, and other structural proteins (*9a, 202*).

Besides the advantages of short elution times, excellent resolution, and high recoveries, a further benefit of these reversed-phase HPLC methods for peptide mapping is the ease of sample preparation. In many cases the crude digest can be loaded directly. The problems associated with overlapping peptides in an otherwise straightforward application can often be remedied by small changes in the pH of the mobile phase or, alternatively, variation in the polarity and concentration of the counterion.

IV. SEPARATION OF PEPTIDES BY GEL PERMEATION, ION-EXCHANGE, AND POLAR ADSORPTION HPLC

A. Introduction

In this section peptide separation by gel permeation, normal-phase adsorption, ion-exchange, and liquid–solid adsorption HPLC will be considered. Compared to reversed-phase HPLC, these elution modes have attracted considerably less interest for peptide separations. This is due to a number of factors arising from the diverse characteristics of the solutes themselves, from the limited suitability of the present range of compatible stationary phases to finely discriminate peptides on the basis of differences in size or charge, and, particularly in the case of separations based on strong ionic interactions, from the practical restraints placed on the choice of composition of mobile phases to adequately manipulate the molecular forces involved in the sorption–desorption events. The current surge of interest in the chemistry of surface bonding of phases onto silica and other types of inorganic support matrices and organic copolymers will undoubtably lead to a new generation of pressure-stable chromatographic adsorbents. These packings should fulfil a similar variety of roles to those long performed by the conventional soft gels in the gel permeation, ion-exchange, and polar adsorption separation of peptides.

B. Gel Permeation and Normal-Phase Liquid–Liquid HPLC of Peptides

Gel permeation chromatography of peptides on soft gels of different porosities has been extensively used over the past three decades. Because selectivity is controlled, in part, by the difference in molecular weight between the solutes [see Eq. (4)], gel permeation systems poorly discriminate peptides of similar size. Traditionally, hydrophilic matrices of low ionic characteristics have been used based on cross-linked polydextran, agarose, polyacrylamide, or glycolmethylacrylate gels. Many of these gels cannot withstand high flow velocities or pressures and undergo deformation with changes in the experimental conditions, e.g., changes in mobile-phase composition or ionic strength, or temperature. Much attention has been recently focused on the use of porous silicas for the high-speed gel filtration of peptides, polypeptides, and proteins. In order to prepare phases suitable for steric exclusion HPLC based on silica support matrices, at least the following four conditions must be met:

(1) The bonded ligand neither selectively repels nor adsorbs amphoteric biomolecules.
(2) The bonded ligand is chemically stable toward buffered solutions (possibly containing denaturants such as 5 *M* guanidine hydrochloride) at least over the range pH 3.0–7.5.
(3) The chemical modification of the silica matrix produces a dense, uniform surface coverage of the ligand.
(4) The ligand must be water wettable but does not act as an ionic or hydrophobic surface.

With present bonding technology, it is not feasible to produce hydrophilic phases which satisfy all these criteria. The strong interactive adsorption of many peptides to naked silica matrices preclude their general use in size exclusion HPLC. Currently, two approaches have been used to reduce the polar adsorption of peptides to silica matrices due to ionic interactions of cationic groups in the solute with charged, nonbridging silanol oxygen atoms and due to hydrogen bonding interactions of free —NH or —OH groups of the peptidic solute with oxygen atoms of the siloxane chains in the matrix. The latter effect can be largely overcome by coating a porous silica with hydrophilic phases such as polyethylene glycols or nonionic detergents. Alternatively, chemical bonding of a suitable hydrophilic phase, such as 1,2-dihydroxy-3-propoxypropylsilyl, to the silica surface can be used to minimize both types of adsorptive effect. A variety of studies have described the use of silica-based hydrophilic supports for the analytical and semipreparative separation of polypeptides and proteins (*203–209*) or the determination of the molecular weight of polypeptides under dissociating conditions (*210, 213*). Despite the obvious potential these supports could offer for the rapid size group separation of peptides, desalting, etc., they have yet to find widespread usage due either to low sample capacity, high cost, the fact that with some supports basic peptides elute in or beyond the inclusion volume, or manufacturing difficulties. Many of the more interesting hydrophilic-phase silica systems for use with peptides of MW 500–5000 are still only available at the laboratory production stage. Encouraging results have been obtained for polypeptide separations on some hydrophilic supports with excellent preparative recoveries of biomolecules in a native form. Gel permeation HPLC will undoubtedly find increasing use for the size fractionation of peptides, particularly with the advent of a wider array of controlled porosities for the silica gel matrix. Porous, hydrophilic silica-based supports now in use cover the pore diameter range from 6 to 100 nm (i.e., from MW 10^3 to 10^6). Column permeability is, however, very dependent on pore volume variations, which limit both resolution and sample capacity. Despite these limitations, the present generation of hydrophilic bonded silicas frequently permit adequate analytical assessment of molecular weight or micropreparative recovery of biological peptides. By coupling gel permeation and reversed-phase HPLC procedures, very rapid analysis of biological extracts can be achieved. Examples of the use of gel permeation HPLC for polypeptide separation include the separation of bovine insulin a-, b-, and c-components under dissociating conditions, with 7 *M* urea–0.1% orthophosphoric acid (*213*), fractionation of the tryptic peptides of human thyroglobulin (*214*) (Fig. 17), complement components (*215*), and neurophysin–vasopressin complexes (*216*). Chemical affinity equilibria, e.g., hydrophobic or ionic interactions, between the peptide and the hydrophilic phase can lead to selectivity effects not in accord with those

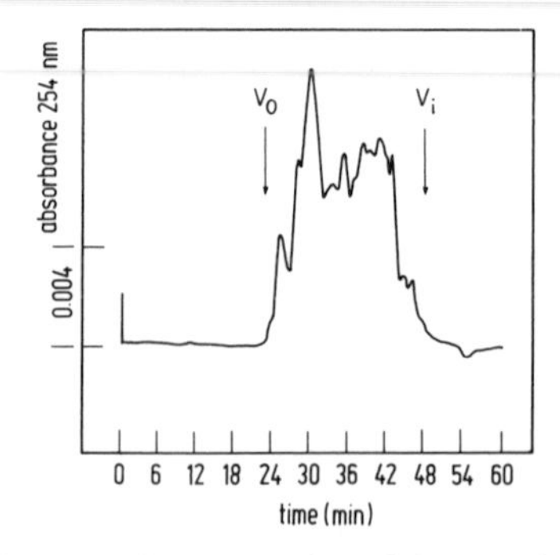

FIG. 17. High-speed gel permeation separation of the tryptic peptides of human thyroglobulin on dual Waters I-125 protein columns using a mobile phase comprised of 7 *M* urea–0.1% H_3PO_4, pH 3.5, at a flow rate of 0.5 ml/min. Reproduced from Hearn *et al.* (*170*).

expected solely on the basis of molecular weight. For example, the separation of insulin (MW ≈ 5630), glucagon (MW ≈ 3550); and somatostatin (MW ≈ 1636) can be achieved on controlled pore glass coated with glyceryl-propylsilane, but these polypeptides elute in reverse order to that anticipated for a true steric exclusion process (*217*). In general, polar interactions with peptides can arise when the ligand is a bonded aminopropyl, cyanopropyl, ether, or glycol moiety. When these interactions dominate the sorption process, retention will follow normal-phase selectivities. Because of their surface properties, the bonded nonpolar and polar stationary phases make excellent supports for classical liquid–liquid partition chromatography. However, because separations can be more easily achieved in the reversed-phase mode, there is little demand for such a use. The improved efficiencies and recoveries for very hydrophobic peptides noted with the use of low concentrations of surface-active alcohols in reversed-phase separations, probably has its origin in the adsorption of the alcohol onto the ligand surface, leading to a dynamic hydrophilic-phase character. The selectivity changes occasionally observed at higher solvent concentrations with acetonitrile and propanol may also arise due to a similar phenomenon. The comparative capabilities of several of the more popular commercially available hydrophilic phases (Waters I series, Toyo Soda TSK-Gel SW series, and LiChrosorb DIOL) for polypeptide separation has been reviewed elsewhere (*11*). Peptides exhibit divergent retention behavior with each of these different phase types, although under appropriately chosen elution conditions, chromatographic behavior approximating a true permeation mode can be achieved (*213, 218, 219*).

Besides the bonded diol or epoxy type of hydrophilic stationary phase, a number of other types of bonded silica supports of varying phase polarities are now available. The use of the nonprotic chemically bonded cyanopropyl silicas for peptide separations has been discussed earlier (p. 102). Mathes and Englehardt have recently introduced (*205*) the 3-acetoamidopropylsilyl phases, and with these supports chemical affinity equilibria between the support and the solute appear to play an insignificant role in the retention process, except for the more basic solutes (*220*). Again these phases are mainly designed for the higher-molecular-weight solutes since they have been based on 100- to 500-Å-pore-diameter silicas. Grafted hydrophilic phases based on bonded chiral tripeptidyl ligands have been used (*79, 221*) to separate isomeric dipeptides, although their column efficiencies were found to be low. Pressure-stable hydrophilic porous copolymer gels with small d_p and appropriate pore size would be a most useful adjunct to their soft gel counterparts. Although a number of porous organic polymer gels, suitable for HPLC procedures, are now available (*14, 16, 57*), their porosity range or surface polarity characteristics are not fully compatible with simple mobile phases for the efficient discrimination of small peptides by size alone. The use of the cross-linked organic polymers, like the Poragels, in the separation of peptides has been reported (*40*). At present, most of these supports show extremely low efficiencies, but current interest in their manufacture will almost certainly result in this situation improving shortly.

C. Ion-Exchange HPLC of Peptides

The fact that in ion-exchange chromatography large selectivity differences can result from small changes in the pH or ionic strength of the eluent and the phase volume [see Eq. (5)], has long been utilized for the separation of amino acids, peptides, and proteins. In the area of high-speed ion-exchange chromatography, ion-exchange resins based on porous, spherical polystyrene–divinylbenzene (PS–DVB) copolymers ($d_p \approx$ 6–12 μm, 8–12% cross-linking) have been available for a number of years. A variety of very reliable automated amino acid analyzers utilize these resins for the analysis of protein hydrolysates where complete resolution of the protein amino acids is possible in <60 min. However, these resins exhibit a low rate of mass exchange (hence the use of elevated temperatures like 60°C to improve efficiencies) and the degree of swelling is affected by the type of buffer, its ionic strength and pH, the temperature, and the extent of cross-linking. Many PS–DVB ion-exchange resins with the low percentage of cross-linkage required for polypeptide separation cannot withstand high flow velocities. These properties have limited the use of PS–DVB ion exchangers to the high-speed separation of amino acids, glycosylated amino acids, and small peptides (*222–227*). Bradshaw

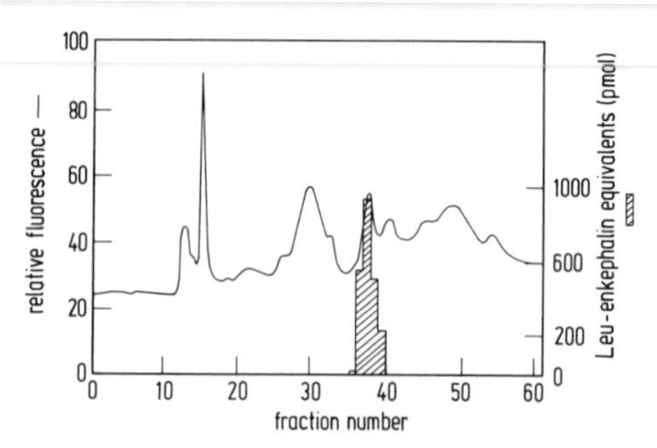

FIG. 18. Chromatography of crude rat β-lipotrophin on a Partisil SCX column (25 × 0.46 cm) using a 100-min linear gradient from 1 *M* pyridine–0.5 *M* acetic acid to 3 *M* pyridine–1.5 *M* acetic acid; flow rate, 50 ml/h; temperature, 50°C. The location of the active fractions is indicated by the hatched region. Reprinted with permission from Rubinstein *et al.* (*134*).

et al. (*228*) have reported a comparative study on globin digests with a variety of cross-linked PS–DVB anion and cation exchangers. Several high-speed analyses of hemoglobin A_{1c} have been based on the weakly acid PS–DVB resins (*229, 230*). With the introduction of the pellicular and, subsequently, the fully porous bonded ion-exchange phases on microparticulate silica gels, the deformation effects noted with the PS–DVB resins have been largely overcome. However, eluent pHs are still generally restricted to below pH 7.5. Improved pH stability should be possible with the fully porous bonded or molded ion exchangers based on alumina or titania. The potential of the strongly anionic or cationic silica-based ion exchangers has been demonstrated (*207, 231, 232*) with proteins. Several reports have described their use for peptide separations with either pyridine–acetic acid buffers or phosphate buffers, e.g., for the isolation of β-lipotropin (*82*) and β-endorphin (*233*) from rat pituitaries (Fig. 18), the separation of Leu- and Met-enkephalin (*234*), and a number of purified peptide hormones (*235*) including nonsuppressible insulin-like activity (NSILA) and parathyroid hormone PTH_{1-34}. See note added in proof, p. 155.

An improved procedure for the preparation of stable anion exchangers on porous microparticulate silicas has been described by Regnier and co-workers (*232*). These and related supports have found application in the profiling of hemoglobin variants and for peptide separation (*236–239*).

D. Liquid–Solid Adsorption HPLC of Peptides

The separation of ionizable compounds by liquid–solid adsorption chromatography has long presented difficulties and the separation of unprotected peptides is no exception. Because of the very polar nature of the basic, acidic, and hydroxylic functional groups in unprotected peptides, attempts to separate peptides on a polar adsorbent surface like silica will result in favorable adsorption isotherms being established with the concomitant displacement of a mobile-phase component such as water or methanol, which was functioning as a protective deactivating layer ensuring homogeneity of the adsorbing sites. In order to achieve desorption, eluents of comparable polarity to the ionogenic solutes are required. Because optimal elution conditions can be difficult to obtain, the outcome is poor resolution, peaks that tail severely and low recoveries. For these reasons, only partially or fully protected peptides can be successfully separated by liquid–solid adsorption HPLC. Mobile-phase selection can be based on elutropic differences, $\Delta\epsilon^\circ$, between solvents and a suitable solvent strength for the isocratic, stepwise, or gradient elution estimated from TLC R_f data. Ionizable solvents at low concentrations (e.g., 0.5% acetic acid) may be required to ensure satisfactory resolution. In these cases, selectivity is very responsive to the protic solvent content and, if gradient elution is used, re-equilibration difficulties may lead to decreased column capacity over a period of time. With microparticulate silica gels with surface areas of 200–500 m^2/g adequate column efficiencies can be obtained with the less polar, fully protected peptides using mobile phases composed of chloroform, an alcohol such as 1- or 2-propanol, ethanol, or methanol, and a low concentration of acetic acid or another weak acid. With the more polar, fully protected peptides or partially protected peptides insoluble in chlorinated hydrocarbons, solvents of greater ϵ° are required. In these circumstances, eluents involving dimethylformamide, hexamethylphosphoramide, or triethylamine are commonly favored. Analytical to large-scale preparative HPLC separations of protected peptides are now possible with the available range of microparticulate silicas. With reasonable selectivity, through outputs of at least 100 g/day are feasible, particularly with the preparative radial compression technology employing high flow rates. Kullman has reported (*240*) the purification of several partially or fully protected precursors of an asymmetrical cystine peptide corresponding to the insulin sequence A_{18-21}-B_{19-26} on Merck silica gel 60 (Fig. 19), using chloroform–methanol eluents. Similar methods have been reported for the purification of fully protected somatostatin and its peptide precursors (*241, 242*) and hydrophobic oligopeptides (*243*). Based on stereochemical arguments similar to those outlined in Section III, I, the separation of diastereoisomeric, fully protected peptides on silica gels is anticipated but, in this case, with the LD-diastereomer eluting more rapidly than the LL-diastereoisomer, i.e., opposite to the reversed-phase elution order. This effect has been observed experimentally (*244,*

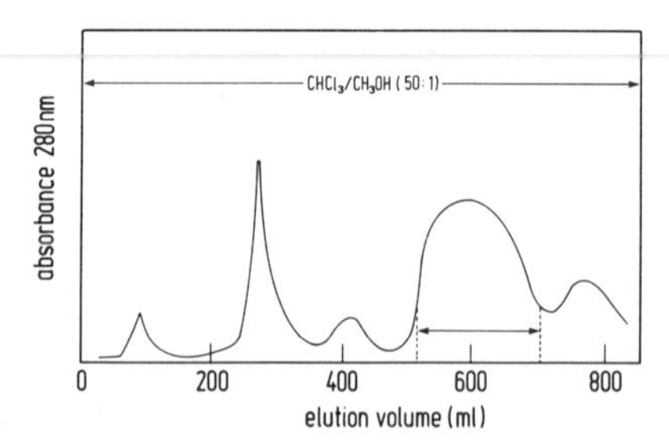

FIG. 19. Chromatography of the crude, protected, asymmetrical cystine insulin peptide ($A_{18-21}-B_{19-26}$) on a Merck silica gel 60 column (47 × 2.3 cm). The crude sample (510 mg) was applied in dimethylformamide (DMF) (1.5 ml) and eluted with $CHCl_3$-MeOH (50 : 1) at a flow rate of 0.5 ml/min. The peak containing the desired peptide (Bpoc-Asn-Tyr-(2,6-Cl_2-Bzl)-Cys-Asn-*O*Bzl-Boc-Cys-Gly-OH) is indicated by an arrow. Reprinted with permission from Kullmann (*190*). Copyright by Marcel Dekker, Inc., New York.

245) and can be utilized as a rapid, alternative method for the assessment of racemization of fully protected peptides during solution–phase synthesis. Finally, ligand-exchange chromatography on a copper–ammonium silica column has been applied to enkephalin degradation studies (*246*) in mouse brain.

V. SUMMARY AND FUTURE PROSPECTS

In this article, the various facets of high-performance liquid–liquid, gel permeation, ion-exchange, and liquid–solid chromatography in the separations of peptides has been critically examined. These methods collectively permit the rapid and selective separation of peptides which are closely related as well as those which differ over a very wide range of polarities and size. A cursory literature search, or a brief examination of the selected bibliography, reveals that by early 1980 reversed-phase HPLC has proved to be the most popular and the most powerful of the modern chromatographic approaches for peptide separation. Reliable microanalytical to large-scale preparative separations have now been developed using chemically bonded reversed phases and a variety of suitably novel mobile phases. Further progress will come from a number of quarters. First, the availability of an improved array of chromatographic supports will extend present capabilities. Of relevance here are the current studies exploring the use of inert supports of more closely defined pore size and surface characteristics. From a practical point of view, we can confidently look forward to support matrices of greater pH and chemical stability. Highly specific stationary phases suitable for high-efficiency bioaffinity separations of peptides should be just one consequence of this current surge of interest in the chemistry of bonded phases (*247*). Second, only a small proportion from the potential range of useful mobile-phase combinations has currently been examined. Additional knowledge on the properties of multicomponent eluents will undoubtedly become available from studies with simple ionogenic substances, but the application of these eluents will greatly simplify the more complex peptide separations. Third, the application of new theoretical concepts will provide the framework for evaluating mobile- and stationary-phase effects and allow reliable indices to be used to predict or rationalize the retention characteristics of a group of peptides. The availability of data banks of hydrophobicity parameters of peptides and surface tension values for multicomponent eluents would be of considerable value in, for example, protein sequence determinations. Improved detection methods will allow greater sensitivities and more selective detection. Postcolumn chemical reactors already fulfill a routine role in fluorescence detection. Shortly, reliable specific detectors, capable of responding to enzymatic, electrochemical, or immunological reactions, will become available (*248*) and their use will simplify the detection of specific peptide components. Clearly such dedicated detectors, in combination with HPLC methods, will have a major impact on the clinical sciences. The recent report (*249*) of a laser fluorescence immunoassay for insulin typifies these developments. Finally we can anticipate the development of reliable automated systems capable of optimizing the elution conditions around the molecular characteristics of the peptides and the sorptive properties of the stationary phase. Fully automated HPLC systems may not necessarily prove to be a general panacea for all areas of peptide separation, but they could certainly provide some enhanced capabilities such as direct, on-line structural mapping of polypeptides and proteins via dedicated mass spectrometers.

Acknowledgments

The support of the National Health and Medical Research Council of Australia and the Medical Research Council of New Zealand is gratefully acknowledged.

REFERENCES

1. J. B. Sumner, *J. Biol. Chem.* **69,** 435 (1926).
2. V. du Vigneaud, C. Ressler, J. M. Swan, C. W. Roberts, P. G. Katsoyannis, and S. Gordon, *J. Am. Chem. Soc.* **75,** 4879 (1953).
3. A. P. Ryle, F. Sanger, L. F. Smith, and R. Kitai, *Biochem. J.* **60,** 541 (1955).

4. J. H. Julliard, T. Shibasaki, N. Ling, and R. Guillemin, *Science* **208,** 183 (1980).
5. R. V. Lewis, A. S. Stern, J. Rossier, S. Stein, and S. Udenfriend, *Biochem. Biophys. Res. Commun.* **89,** 822 (1979).
6. M. T. W. Hearn, and A. J. Patterson, submitted for publication.
7. J. Rivier, R. Kaiser, and R. Galyean, *Biopolymers* **17,** 1927 (1978).
8. M. T. W. Hearn, C. A. Bishop, W. S. Hancock, D. R. K. Harding, and G. D. Reynolds, *J. Liq. Chromatogr.* **2,** 1 (1979).
9. K. Krummen, *J. Liq. Chromatogr.* **3,** 1243 (1980).
9a. M. T. W. Hearn, F. E. Regnier, and C. T. Wehr, "Proceedings of the First International Symposium on HPLC of Proteins and Peptides." Academic Press, New York, 1983.
9b. M. T. W. Hearn, F. E. Regnier, K. Unger, and C. T. Wehr, "Proceedings of the Second International Symposium of Proteins, Peptides and Polynucleotides" Elsevier-North Holland Publishing Co., Amsterdam, 1983.
10. M. T. W. Hearn, F. E. Regnier, and C. T. Wehr, *Am. Lab.* **14,** 18 (1982).
11. M. T. W. Hearn, *in* "Advances in Chromatography" (J. C. Giddings, P. Brown, and J. Cazes, eds.). Dekker, New York, **20,** 1, 1982.
12. J. Sparrow and W. S. Hancock, this volume.
13. M. Rubinstein, *Anal. Chem.* **98,** 1 (1979).
13a. M. T. W. Hearn, *in* "Methods in Enzymology" (W. B. Jacoby, ed), Vol 22 supplement, Academic Press, New York, in press.
13b. M. T. W. Hearn and W. S. Hancock, *in* "Biological/Biomedical Applications of Liquid Chromatogranhy" (G. L. Hawk, ed), Marcel Dekker, New York, p 243, 1979.
14. R. E. Majors, *J. Chromatogr. Sci.* **15,** 334 (1977).
15. K. K. Unger, "Porous Silica: Its Properties and Use as Support in Column Liquid Chromatography." Elsevier, Amsterdam, 1978.
16. J. H. Knox, J. N. Done, A. F. Fell, M. T. Gilbert, A. Pryde, and R. A. Wall, "High-Performance Liquid Chromatography." Edinburgh Univ. Press, Edinburgh, 1979.
17. M. J. O'Hare and E. C. Nice, *J. Chromatogr.* **171,** 209 (1979).
17a. B. Grego and M. T. W. Hearn, *Chromatographia* **14,** 589 (1981).
17b. M. T. W. Hearn and B. Grego, *J. Chromatogr.* **218,** 497 (1981).
17c. M. T. W. Hearn and B. Grego, *J. Chromatogr.*, **255,** 125 (1983).
18. J. Poráth, L. Sundberg, N. Fornstedt, and I. Olsson, *Nature* (*London*) **245,** 465 (1973).
19. S. Shaltiel, *in* "Methods in Enzymology (W. B. Jakoby and M. Wilchek, eds.), Vol. 34, p. 126. Academic Press, New York, 1974.
20. A. H. Nishikawa and P. Bailon, *Anal. Biochem.* **68,** 274 (1975).
21. M. T. W. Hearn, G. S. Bethell, J. S. Ayers, and W. S. Hancock, *J. Chromatogr.* **185,** 463 (1979).
21a. M. T. W. Hearn, E. J. Harris, J. F. Bethell, W. S. Hancock, and J. A. Ayers, *J. Chromatogr.* **218,** 483 (1981).
21b. J. Bethell, J. A. Ayers, W. S. Hancock, and M. T. W. Hearn, *J. Chromatogr.* **219,** 361 (1981).
22. A. J. P. Martin and R. L. M. Synge, *Biochem. J.* **35,** 1358 (1941).
23. D. Yamashiro, *Nature* (*London*) **201,** 76 (1964).
24. J. G. R. Hurrell, R. J. Fleming, and M. T. W. Hearn, *J. Liq. Chromatogr.* **3,** 473 (1980).
25. C. A. Meyers, D. H. Coy, W. Y. Huang, A. V. Schally, and T. W. Redding, *Biochemistry* **17,** 2326 (1978).
26. D. Yamashiro, *Int. J. Pept. Protein Res.* **13,** 5 (1979).
27. W. S. Hancock, C. A. Bishop, and M. T. W. Hearn, *FEBS Lett.* **72,** 139 (1976).
27a. R. Burgus and J. Rivier, *Pept. Proc. Eur. Pep. Symp. 14th, 1976* p. 85 (1976).
28. C. Horváth and W. Melander, "High-Performance Liquid Chromatography: Ad-

vances and Perspectives" (C. Horvath, ed.), Vol. 1, pp. 113–319. Academic Press, New York, 1980.
29. M. T. W. Hearn, B. Grego, and W. S. Hancock, *J. Chromatogr.* **185,** 429 (1979).
30. M. T. W. Hearn and B. Grego, *J. Chromatogr.* (in press).
31. C. M. Riley, E. Tomlinson, and T. M. Jefferies, *J. Chromatogr.* **185,** 197 (1979).
32. R. F. Rekker, "The Hydrophobic Fragmental Constant." Elsevier, Amsterdam, 1977.
33. I. Molnar and C. Horváth, *J. Chromatogr.* **142,** 623 (1977).
34. M. T. W. Hearn and B. Grego, submitted for publication.
35. M. T. W. Hearn and W. S. Hancock, *J. Chromatogr. Sci.* **10,** 288 (1980).
36. C. Horvath, W. Melander, and I. Molnar, *J. Chromatogr.* **125,** 129 (1976).
37. O. Sinanoglu, *in* "Molecular Associations in Biology" (B. Pullman, ed.), p. 427. Academic Press, New York, 1968.
38. T. Halicioglu and O. Sinanoglu, *Ann. N.Y. Acad. Sci.* **158,** 308 (1969).
39. C. H. Lochmuller and D. R. Wilder, *J. Chromatogr. Sci.* **17,** 574 (1979).
39a. M. T. W. Hearn and B. Grego, in preparation.
39b. M. T. W. Hearn and B. Grego, submitted for publication.
39c. J. D. Pearson, N. T. Lin, and F. E. Regnier, *Anal. Biochem.,* **124,** 217 (1982).
40. M. T. W. Hearn and B. Grego, in preparation.
41. J. J. Hansen, T. Greibrokk, B. L. Currie, K. N. Johansson, and K. Folkers, *J. Chromatogr.* **135,** 155 (1977).
42. E. Lundanes and T. Greibrook, *J. Chromatogr.* **149,** 241 (1978).
43. R. V. Lewis, A. Fallon, S. Stein, K. D. Gibson, and S. Udenfriend, *Anal. Biochem.* **103** (in press).
44. B. L. Karger, K. A. Cohen, D. Schellenberg, B. Grego, and M. T. W. Hearn, submitted for publication.
45. E. P. Kroeff and D. J. Pietrzyk, *Anal. Chem.* **50,** 502 (1978).
45a. S. J. Su, B. Grego, B. Niven, and M. T. W. Hearn, *J. Liq. Chromatogr.* **4,** 1745 (1981).
46. E. P. Kroeff and D. J. Pietrzyk, *Anal. Chem.* **50,** 1353 (1978).
46a. J. L. Meek, *Proc. Natl. Acad. Sci. U.S.A.* **77,** 1632 (1980).
47. K. Krummen and R. W. Frei, *J. Chromatogr.* **132,** 27 (1977).
48. J. E. Rivier, L. H. Lazarus, M. H. Perrin, and M. R. Brown, *J. Med. Chem.* **20,** 1409 (1977).
49. J. E. Rivier and M. R. Brown, *Biochemistry* **17,** 1766 (1978).
50. S. Terabe, R. Konaka, and K. Inouye, *J. Chromatogr.* **172,** 163 (1979).
51. Y. Nozaki and C. Tanford, *J. Biol. Chem.* **246,** 2211 (1971).
52. A. Leo, C. Hansch, and D. Elkins, *Chem. Rev.* **71,** 525 (1971).
52a. P. Y. Chou and G. D. Fasman, *Biochemistry* **13,** 211 (1974).
52b. J. L. Meek and Z. Rosetti, *J. Chromatogr.* **211,** 15 (1981).
52c. K. J. Wilson, A. Honnegger, R. P. Stotzel, and G. J. Hughes, *Biochem. J.* **199,** 31 (1981).
52d. H. J. P. Bennett, C. A. Browne, and S. Solomon, *Anal. Biochem.* **124,** 201 (1982).
52e. J. J. L'Italien and R. A. Laursen, *J. Biol. Chem.* **256,** 8092 (1981).
52f. M. T. W. Hearn, B. Grego and F. Lambrou, submitted for publication.
53. B. Larsen, B. L. Fox, M. F. Burke, and V. J. Hruby, *Int. J. Pept. Protein Res.* **13,** 12 (1979).
54. P. J. Schoenmakers, H. A. H. Billiet, and L. D. Galan, *J. Chromatogr.* **185,** 179 (1979).
54a. M. T. W. Hearn and B. Grego, *J. Liq. Chromatogr.* (in press).
54b. K. E. Bij, C. Horváth, W. R. Melander, and A. Nahum, *J. Chromatogr.* **203,** 65 (1981).
55. C. A. Bishop, W. S. Hancock, S. O. Brennan, R. W. Carrell, and M. T. W. Hearn, *J. Liq. Chromatogr.* **4,** 599 (1981).

56. R. P. W. Scott and P. Kucera, *J. Chromatogr.* **175,** 51 (1979).
57. S. Stein, *in* "The Peptides" (E. Gross, ed.). Academic Press, New York p. 73, 1981.
58. S. N. Deming and M. L. H. Turoff, *Anal. Chem.* **50,** 546 (1978).
59. J. C. Kraak, K. M. Jonker, and J. F. K. Huber, *J. Chromatogr.* **142,** 671 (1977).
60. J. E. Rivier, *J. Liq. Chromatogr.* **1,** 343 (1978).
61. D. J. Pietrzyk and C. H. Chu, *Anal. Chem.* **49,** 860 (1977).
62. C. Horváth, W. Melander, I. Molnar, and P. Molnar, *Anal. Chem.* **49,** 2295 (1977).
63. W. S. Hancock, C. A. Bishop, R. L. Prestidge, D. R. K. Harding, and M. T. W. Hearn, *Science* **200,** 1168 (1978).
64. W. S. Hancock, C. A. Bishop, L. J. Meyer, D. R. K. Harding, and M. T. W. Hearn, *J. Chromatogr.* **161,** 291 (1978).
65. W. S. Hancock, C. A. Bishop, J. E. Battersby, D. R. K. Harding, and M. T. W. Hearn, *J. Chromatogr.* **168,** 377 (1979).
66. M. T. W. Hearn and W. S. Hancock, *Trends Biochem. Sci.* **4,** 58 (1979).
67. W. S. Hancock, C. A. Bishop, and M. T. W. Hearn, *Chem. N.Z.* **43,** 11 (1979).
68. M. T. W. Hearn, *Adv. Chromatogr.* **18,** 59 (1980).
68a. J. H. Knox and R. A. Hartwick, *J. Chromatogr.* **204,** 3 (1981).
69. R. S. Deelder, H. A. J. Linssen, A. P. Konijnendijk, and J. L. M. Van de Venne, *J. Chromatogr.* **185,** 241 (1979).
70. M. T. W. Hearn, S. J. Su, and B. Grego, *J. Liq. Chromatogr.* **4,** 1547 (1981).
71. W. S. Hancock, C. A. Bishop, R. L. Prestidge, D. R. K. Harding, and M. T. W. Hearn, *J. Chromatogr.* **153,** 391 (1978).
72. M. T. W. Hearn, W. S. Hancock, J. G. R. Hurrell, R. J. Fleming, and B. Kemp, *J. Liq. Chromatogr.* **2,** 919 (1979).
73. K. Tsuji and J. H. Robertson, *J. Chromatogr.* **112,** 663 (1975).
74. M. T. W. Hearn, *J. Liq. Chromatogr.* **3,** 1255, 1980.
75. C. S. Fullmer and R. H. Wasserman, *J. Biol. Chem.* **254,** 7208 (1979).
76. M. T. W. Hearn and B. Grego, unpublished observations.
77. C. Schwabe and J. K. McDonald, *Biochem. Biophys. Res. Commun.* **74,** 1501 (1977).
78. G. E. Gerber, R. J. Anderegg, W. C. Herlihy, C. P. Gray, K. Biemann, and H. G. Khorana, *Proc. Natl. Acad. Sci. U.S.A.* **76,** 227 (1979).
79. E. J. Kikta and E. Grushka, *J. Chromatogr.* **135,** 367 (1977).
80. B. Larsen, V. Viswanatha, S. Y. Chang, and V. J. Hruby, *J. Chromatogr. Sci.* **16,** 207 (1978).
81. M. Rubinstein, S. Rubinstein, P. C. Familletti, R. S. Miller, A. A. Waldman, and S. Pestka, *Proc. Natl. Acad. Sci. U.S.A.* **76,** 640 (1979).
82. M. Rubinstein, S. Stein, L. D. Gerber, and S. Udenfriend, *Proc. Natl. Acad. Sci. U.S.A.* **74,** 3052 (1977).
83. C. E. Dunlap, S. Gentleman, and L. I. Lowney, *J. Chromatogr.* **160,** 191 (1978).
84. C. McMartin and G. E. Purdon, *J. Endocrinol.* **77,** 67 (1978).
85. C. A. Bishop, D. R. K. Harding, L. J. Meyer, W. S. Hancock, and M. T. W. Hearn, *J. Chromatogr.* **192,** 222 (1980).
86. H. P. J. Bennett, C. A. Browne, and S. Solomon, *J. Liq. Chromatogr.* **3,** 1353 (1980).
87. W. A. Schroeder, J. B. Shelton, J. R. Shelton, and D. Powars, *J. Chromatogr.* **174,** 385 (1979).
88. M. T. W. Hearn, J. F. Cutfield, and L. R. McInnes, submitted for publication.
89. D. H. Coy, P. Gill, A. J. Kastin, A. Dupont, L. Cusan, F. Labrie, D. Britton, and R. Fertel, *Pept. Proc. Am. Pept. Symp. 5th, 1977* p. 107 (1977).
90. B. L. Karger, W. S. Wong, R. L. Viavattene, J. N. Le Page, and G. Davies, *J. Chromatogr.* **167,** 253 (1978).
91. M. Rubinstein, S. Chen Kiang, S. Stein, and S. Udenfriend, *Anal. Biochem.* **95,** 117 (1979).

92. S. Kimura, R. V. Lewis, L. D. Gerber, L. Brink, M. Rubinstein, S. Stein, and S. Udenfriend, *Proc. Natl. Acad. Sci. U.S.A.* **76,** 1756 (1979).
93. J. A. Feldman, M. L. Cohn, and D. Blair, *J. Liq. Chromatogr.* **1,** 833 (1978).
94. J. Spiess, J. E. Rivier, J. A. Rodkey, C. D. Bennett, and W. Vale, *Proc. Natl. Acad. Sci. U.S.A.* **76,** 2974 (1979).
95. M. E. F. Biemond, W. A. Sipman, and J. Olivie, *J. Liq. Chromatogr.* **2,** 1407 (1979).
96. W. L. Hollaway, A. S. Bhown, J. E. Mole, and J. C. Bennet, *Chromatogr. Sci.* **10,** 163 (1979).
97. H. P. J. Bennett, C. A. Browne, and S. Solomon, *Biochemistry* **20,** 4530 (1981).
98. J. D. Pearson, W. C. Mahoney, M. A. Hermodson, and F. E. Regnier, *J. Chromatogr.* **207,** 325 (1981).
99. D. Robinson, H. G. Burger, J. C. Findlay, F. J. Morgan, and M. T. W. Hearn, submitted for publication.
100. W. C. Mahoney and M. A. Hermodson, *J. Biol. Chem.* **255,** 11199 (1980).
101. M. T. W. Hearn and A. J. Paterson, *J. Liq. Chromatogr.*, in press.
102. W. J. Kohr, R. Keck, and R. N. Harkins, *Anal. Biochem.* **122,** 348, 1982.
103. C. A. Bishop, D. R. K. Harding, L. J. Meyer, W. S. Hancock, and M. T. W. Hearn, *J. Liq. Chromatogr.* **4,** 661, 1981.
104. D. R. K. Harding, C. A. Bishop, M. F. Tarttelin, and W. S. Hancock, *Int. J. Pept. Protein Res.* **18,** 214 (1981).
105. L. R. Snyder, *J. Chromatogr.* **179,** 167 (1979).
106. R. Lemieux and A. Barton, *Can. J. Chem.* **49,** 767 (1971).
107. E. Ellenbogen, *J. Am. Chem. Soc.* **74,** 5198 (1952).
108. P. Flory and P. Schimmel, *J. Am. Chem. Soc.* **89,** 6807 (1967).
109. J. Rivier and R. Burgus, *Chromatogr. Sci.* **10,** 147 (1979).
110. W. R. Cahill, E. P. Kroeff, and D. J. Pietrzyk, *J. Liq. Chromatogr.* **3,** 1319, (1980).
111. B. Larsen, B. L. Fox, M. F. Burke, and V. J. Hruby, *Int. J. Pept. Protein Res.* **13,** 12 (1979).
112. C. Hunter, K. Sugden, and J. G. Lloyd-Jones, *J. Liq. Chromatogr.* **3,** 1335, (1980).
113. D. H. Coy, *Chromatogr. Sci.* **10,** 283 (1979).
114. J. E. Rivier, J. Spiess, M. Perrin, and W. Vale, *Chromatogr. Sci.* **10,** 223 (1979).
115. J. M. Manning and S. Moore, *J. Biol. Chem.* **243,** 5591 (1968).
116. J. H. Knox and G. Szokan, *J. Chromatogr.* **171,** 439 (1979).
117. F. Nachtmann, *J. Chromatogr.* **176,** 391 (1979).
118. S. I. Sallay and S. Oroszlan, *Chromatogr. Sci.* **10,** 199 (1979).
119. F. Maxl and K. Krummen, *Pharm. Acta Helv.* **53,** 207 (1978).
120. K. Krummen, F. Maxl, and F. Nachtmann, *Pharm. Technol.* **3,** 77 (1979).
121. J. Rivier, R. Wolbers, and R. Burgus, *Pept. Proc. Am. Pept. Symp., 5th, 1977* p. 52 (1977).
122. P. Rivaille, D. Raulais, and G. Milhand, *J. Chromatogr. Sci.* **12,** 273 (1979).
123. U. Damgaard and J. Markussen, *Horm. Metab. Res.* **11,** 580 (1979).
124. A. Dinner and L. Lorenz, *Anal. Chem.* **51,** 1872 (1979).
125. D. Bataille, J. Besson, C. Gespach, and G. Rosselin, *Horm. Recept. Dig. Nutr., Proc. Int. Symp. Horm. Recept. Dig. Tract Physiol., 2nd, 1979* p. 79 (1979).
126. E. C. Nice and M. J. O'Hare, *J. Chromatogr.* **162,** 401 (1979).
127. R. A. Martinelli and H. A. Scheraga, *Anal. Biochem.* **96,** 246 (1979).
128. E. C. Nice, M. Capp, and M. J. O'Hare, *J. Chromatogr.* **185,** 413 (1979).
129. A. S. Stern, R. V. Lewis, S. Kimura, J. Rossier, L. D. Gerber, L. Brink, S. Stein, and S. Udenfriend, *Proc. Natl. Acad. Sci. U.S.A.* **76,** 6680 (1979).
130. J. G. Loeber, J. Verhoef, J. P. H. Burbach, and A. Witter, *Biochem. Biophys. Res. Commun.* **86,** 1288 (1979).

131. W. Y. Huang, R. C. C. Chang, A. J. Kastin, D. H. Coy, and A. V. Schally, *Proc. Natl. Acad. Sci. U.S.A.* **76,** 6177 (1979).
132. D. H. Live, W. C. Agosta, and D. Cowburn, *J. Org. Chem.* **42,** 3556 (1977).
132a. J. T. Russell, M. J. Brownstein, and H. Gainer, *Proc. Natl. Acad. Sci. U.S.A.* **76,** 6086 (1979).
133. J. A. Glasel, *J. Chromatogr.* **145,** 469 (1978).
134. W. Monch and W. Dehnen, *J. Chromatogr.* **147,** 415 (1978).
135. G. W. K. Fong and B. T. Kho, *J. Liq. Chromatogr.* **2,** 957 (1979).
136. D. Rudman, R. K. Chawla, and B. M. Hollins, *J. Biol. Chem.* **254,** 1010 (1979).
137. P. Crine, F. Cossard, N. G. Seidah, L. Blanchette, M. Lis, and M. Chrétien, *Proc. Natl. Acad. Sci. U.S.A.* **76,** 5085 (1979).
138. C. Gianoulakis, N. G. Seidah, R. Roughier, and M. Chrétein, *J. Biol. Chem.* **254,** 11903 (1979).
139. W. Monch and W. Dehnen, *J. Chromatogr.* **140,** 260 (1977).
140. A. Harmar, J. G. Schofield, and P. Keen, *Nature (London)* **284,** 267 (1980).
141. J. E. Rivier, J. Spiess, M. Perrin, and W. Vale, *J. Chromatogr. Sci.* **12,** 223 (1979).
142. H. P. J. Bennett, A. M. Hudson, C. McMartin, and G. E. Purdon, *Biochem. J.* **168,** 9 (1977).
143. D. H. Coy, *J. Chromatogr. Sci.* **12,** 283 (1979).
144. N. Ling, *Biochem. Biophys. Res. Commun.* **74,** 248 (1977).
145. A. Goldstein, S. Tachibana, L. I. Lowney, M. Hunkapiller, and L. Hood, *Proc. Natl. Acad. Sci. U.S.A.* **76,** 6666 (1979).
146. R. V. Lewis, S. Stein, and S. Udenfriend, *Int. J. Pept. Protein Res.* **13,** 493 (1979).
147. S. Gentleman, L. I. Lowney, B. M. Cox, and A. Goldstein, *J. Chromatogr.* **153,** 274 (1978).
148. M. C. Beinfeld, R. T. Jensen, and M. J. Brownstein, *J. Liq. Chromatogr.* **3,** 1299, (1980).
149. D. V. Goeddel, D. G. Kleid, F. Bolivar, H. L. Heyneker, D. G. Yansura, R. Crea, T. Hirose, A. Kraszewski, K. Itakura, and A. D. Riggs, *Proc. Natl. Acad. Sci. U.S.A.* **76,** 106 (1979).
150. F. Nachtmann and K. Gstrein, personal communication.
151. J. A. Clements, J. W. Funder, K. Tracy, F. J. Morgan, D. J. Campbell, P. Lewis, and M. T. W. Hearn, *Endocrinology* (in press).
152. L. F. Congote, H. P. J. Bennett, and S. Solomon, *Biochem. Biophys. Res. Commun.* **89,** 851 (1979).
153. W. M. M. Schaaper, D. Voskamp, and C. Olieman, *J. Chromatogr.* **195,** 181 (1980).
154. T. L. O'Donohue, C. G. Charlton, R. L. Miller, G. Boden, and D. M. Jacobowitz, *Proc. Natl. Acad. Sci. U.S.A.* **78,** 5221 (1981).
155. M. Carlquist, H. Jornvall, and V. Mutt, *FEBS Lett.* **127,** 71 (1981).
156. J. F. M. Kinkel, G. Heuver, and J. C. Kraak, *Chromatographia* **13,** 145 (1980).
157. J. E. Smart, H. Oppermann, A. P. Czernilofsky, A. F. Purchio, R. L. Erikson, and J. M. Bishop, *Proc. Natl. Acad. Sci. U.S.A.* **78,** 6013 (1981).
158. H. P. J. Bennett, C. A. Browne, and S. Solomon, *Proc. Natl. Acad. Sci. U.S.A.* **78,** 4713 (1981).
159. J. Spiess, J. Rivier, C. Rivier, and W. Vale, *Proc. Natl. Acad. Sci. U.S.A.* **78,** 6517 (1981).
160. A. B. Roberts, M. A. Anzano, L. C. Lamb, J. M. Smith, and M. B. Sporu, *Proc. Natl. Acad. Sci. U.S.A.* **78,** 5339 (1981).
161. A. B. Roberts, M. A. Anzano, L. C. Lamb, J. M. Smith, C. A. Frolik, H. Marquardt, G. J. Todaro, and M. B. Sporn, *Nature (London)* **295,** 417 (1982).
162. M. T. W. Hearn and B. Grego, submitted for publication.

163. M. T. W. Hearn, B. Grego, and G. E. Chapman, *J. Liq. Chromatogr.* (in press).
164. I. M. Chaiken and C. J. Hough, *Anal. Biochem.* **107,** 11 (1980).
165. W. Santi, J. M. Huen, and R. W. Frei, *J. Chromatogr.* **115,** 423 (1975).
166. D. D. Gay and R. A. Lahti, *Int. J. Pept. Protein Res.* **18,** 107 (1981).
167. J. R. Walsh and H. D. Niall, *Endocrinology* **107,** 1258 (1980).
168. H. P. J. Bennett, A. M. Hudson, L. Kelly, C. McMartin, and G. E. Purdon, *Biochem. J.* **175,** 1139 (1978).
169. M. J. Spiro, *J. Biol. Chem.* **245,** 5820 (1970).
170. W. S. Hancock, C. A. Bishop, R. L. Prestidge, and M. T. W. Hearn, *Anal. Biochem.* **89,** 203 (1978).
171. M. T. W. Hearn, B. Grego, and P. Stanton, in preparation.
172. W. L. Hollaway, R. L. Prestidge, A. S. Bhown, J. E. Mole, and J. C. Bennett, *in* "Advances in Chromatography and Electrophoresis" (A. Frigerio, ed.). Elsevier/North-Holland Publ., Amsterdam (in press).
173. M. T. W. Hearn, B. Grego, A. Hobbs, and M. Smith, *J. Liq. Chromatogr.* **4,** 651 (1981).
174. R. W. Compans, K. Nakamura, M. G. Roth, W. L. Hollaway, and M. C. Kemp, *Proc. Int. Influenza Symp.* (in press).
175. M. McMillan, J. M. Cecka, L. Hood, D. B. Murphy, and H. O. McDevitt, *Nature (London)* **277, 663** (1979).
176. R. G. Cook, J. D. Capra, J. L. Bednarczyk, J. W. Uhr, and E. S. Vitetta, *J. Immunol.* **123,** 2799 (1979).
177. R. L. Jilka and S. Pestka, *J. Biol. Chem.* **254,** 9270 (1979).
178. G. J. Hughes, K. Winterhalter, and K. J. Wilson, *FEBS Lett.* **108,** 81 (1979).
179. T. A. Stoming, F. A. Garver, M. A. Gangarosa, J. M. Harrison, and T. H. J. Huisman, *Anal. Biochem.* **96,** 113 (1979).
180. G. D. Efremov, J. B. Wilson, and T. H. J. Huisman, *Biochim. Biophys. Acta* **579,** 421 (1979).
181. J. B. Wilson, H. Lam, P. Pravatmuang, and T. H. J. Huisman, *J. Chromatogr.* **179,** 271 (1979).
182. E. Minasian, S. Sharma, S. Leach, B. Grego, and M. T. W. Hearn, *Anal. Biochem.* (in press).
183. W. S. Hancock, J. D. Capra, W. A. Bradley, and J. T. Sparrow, *J. Chromatogr.* **206,** 59 (1981).
184. M. T. W. Hearn, B. Grego, and C. A. Bishop, *J. Liq. Chromatogr.* **4,** 1725 (1981).
185. P. Bohlen and G. Kleeman, *J. Chromatogr.* **205,** 65 (1981).
186. G. J. Hughes, C. de Jong, R. W. Fischer, K. H. Winterhalter, and K. J. Wilson, *Biochem. J.* **199,** 61 (1981).
187. C. Gerard and T. E. Hugli, *J. Biol. Chem.* **255,** 4710 (1980).
188. E. Heller, I. K. Kaczmarck, M. W. Hunkapiller, L. E. Hood, and F. Strumwasser, *Proc. Natl. Acad. Sci. U.S.A.* **77,** 2328 (1980).
189. S. Kimura, R. V. Lewis, A. S. Stein, J. Rossier, S. Stein, and S. Udenfriend, *Proc. Natl. Acad. Sci. U.S.A.* **77,** 1681 (1980).
190. N. G. Seidah, R. Routhier, S. Benjannet, N. Lariviere, F. Gossard, and M. Chrétien, *J. Chromatogr.* **193,** 291 (1980).
191. A. A. Kuma, J. H. Mangum, D. J. Blankenship, and J. H. Freisheim, *J. Biol. Chem.* **256,** 8870 (1981).
192. F. A. Robey and T. Y. Liu, *J. Biol. Chem.* **256,** 969 (1981).
193. M. W. Pierce, J. L. Palmer, H. T. Kentman, and J. Avruch, *J. Biol. Chem.* **256,** 8867 (1981).
194. S. S. Dignam and P. Setlow, *J. Biol. Chem.* **255,** 8408 (1980).
195. A. R. Kerlavage and S. Taylor, *J. Biol. Chem.* **255,** 8483 (1980).

196. N. C. Nelson and S. Taylor, *J. Biol. Chem.* **256,** 3743 (1981).
197. W. P. Levy, M. Rubinstein, J. Shively, U. D. Valle, C. V. Lai, J. Moschera, L. Brink, L. Gerber, S. Stein, and S. Petska, *Proc. Natl. Acad. Sci. U.S.A.* **78,** 6186 (1981).
198. J. B. Howard, *J. Biol. Chem.* **255,** 7082 (1980).
198a. R. P. Swenson and J. B. Howard, *J. Biol. Chem.* **255,** 8087 (1980).
199. M. T. W. Hearn, P. A. Smith, and C. Mallia, *Biosci. Rep.* **2,** 247 (1982).
200. D. J. Lundell, R. C. Williams, and A. N. Glazer, *J. Biol. Chem.* **256,** 3586 (1981).
201. F. S. Esch, P. Bohlen, A. S. Otsuka, M. Yoshida, and W. S. Allison, *J. Biol. Chem.* **256,** 9054 (1981).
202. P. K. Rhodes, K. Gibson, and J. Miller, *Biochemistry* **20,** 3117 (1980).
203. R. Roumeliotis and K. K. Unger, *J. Chromatogr.* **185,** 445 (1979).
204. D. E. Schmidt, R. W. Giese, D. Conron, and B. L. Karger, *Anal. Chem.* **52,** 177 (1980).
205. D. Mathes and H. Engelhardt, *Naturwissenschaften* **66,** 51 (1979).
206. M. T. W. Hearn, B. Grego, C. A. Bishop, and W. S. Hancock, *J. Liq. Chromatogr.* **3,** 1549 (1980).
207. S. H. Chang, K. M. Gooding, and F. E. Regnier, *J. Chromatogr.* **125,** 103 (1976).
208. S. Rokushika, T. Ohkawa, and H. Hatano, *J. Chromatogr.* **176,** 456 (1979).
209. T. Hashimoto, H. Sasaki, M. Aiura, and Y. Kato, *J. Chromatogr.* **160,** 301 (1978).
210. K. Fukano, K. Komuja, H. Sasaki, and T. Hashimoto, *J. Chromatogr.* **166,** 47 (1978).
211. N. Ui, *Anal. Biochem.* **97,** 65 (1979).
212. T. Imamura, K. Konishi, M. Yokoyama, and K. Konishi, *Biochem. J.* **86,** 639 (1979).
213. B. S. Welinder, *J. Liq. Chromatogr.* **3,** 1399 (1980).
214. M. T. W. Hearn, A. J. Paterson, and B. Grego, submitted for publication.
215. M. A. Niemann, W. L. Hollaway, and J. E. Mole, *HRC CC, J. High. Resolut. Chromatogr. Chrmatogr. Commun.* **2,** 743 (1979).
216. K. A. Gruber, J. M. Whitaker, and M. Morris, *Anal. Biochem.* **97,** 176 (1979).
217. L. J. Fisher, R. L. Theiss, and D. Charllowski, *Anal. Chem.* **50,** 2143 (1978).
218. J. E. Rivier, *J. Chromatogr.* **202,** 271 (1980).
219. M. E. Himmel and P. G. Squire, *Int. J. Pept. Protein Res.* **17,** 365 (1981).
220. H. Engelhardt, G. Uhr, and M. T. W. Hearn, *J. Liq. Chromatogr.* **4,** 1361 (1981).
221. G. W. K. Fong and E. Grushka, *J. Chromatogr.* **142,** 299 (1977).
222. Z. Deyl, K. Macek, and J. Janak, "Column Liquid Chromatography." Elsevier, Amsterdam, 1975.
223. M. M. Tikhomirov, A. Y. Khorlin, W. Voelter, and H. Bauer, *J. Chromatogr.* **167,** 197 (1978).
224. W. Voelter and K. Zech, *J. Chromatogr.* **112,** 643 (1975).
225. W. Voelter, H. Bauer, S. Fuchs, and E. Pietrzik, *J. Chromatogr.* **153,** 433 (1978).
226. J. R. Cronin, S. Pizzarellow, and W. E. Gandy, *Anal. Biochem.* **93,** 174 (1979).
227. M. vander Rest, W. G. Cole, and F. H. Glorieux, *Biochem. J.* **161,** 527 (1977).
228. R. A. Bradshaw, O. J. Bates, and J. R. Benson, *J. Chromatogr.* **187,** 27 (1980).
229. J. E. Davis, J. M. McDonald, and L. Jarett, *Diabetes* **27,** 102 (1978).
230. R. A. Cole, J. S. Soelder, P. J. Dunn, and H. F. Dunn, *Metab., Clin. Exp.* **27,** 289 (1978).
231. S. H. Chang, R. Noel, and F. E. Regnier, *Anal. Chem.* **48,** 1839 (1976).
232. A. J. Alpert and F. E. Regnier, *J. Chromatogr.* **185,** 375 (1979).
233. M. Rubinstein, S. Stein, and S. Udenfriend, *Proc. Natl. Acad. Sci. U.S.A.* **74,** 4969 (1977).
234. J. L. Meek and T. P. Bohan, *Adv. Biochem. Psychopharmacol.* **19,** 141 (1978).
235. A. N. Radhakrishnan, S. Stein, A. Licht, K. A. Gruber, and S. Udenfriend, *J. Chromatogr.* **132,** 552 (1977).
235a. T. Isobe, T. Takayasu, N. Takai, and T. Okuyama, *Anal. Biochem.,* **122,** 417 (1982).

235b. L. Sodenberg, L. Wahlstrom, and J. Bergstrom, *Protides of the Biological Fluids*, **30** (in press).
235c. P. G. Stanton, R. J. Simpson, F. Lambrou, and M. T. W. Hearn, *J. Chromatogr.* (in press).
235d. M. T. W. Hearn and D. J. Lyttle, *J. Chromatogr.* **218,** 497 (1981).
235e. L. A. A. E. Slayterman and J. Wijdenes, *J. Chromatogr.* **150,** 31 (1978).
235f. J. P. Emond and M. Page, *J. Chromatogr.,* **200,** 57, (1980).
236. M. Dizdoroglu and M. G. Simic, *J. Chromatogr.* **195,** 119 (1980).
237. K. M. Gooding, K. Lu, and F. E. Regnier, *J. Chromatogr.* **164,** 506 (1979).
238. N. Takahashi, T. Isobe, H. Kasai, K. Seta, and T. Okuyama, *Anal. Biochem.* **115,** 181 (1981).
239. H. Mabuchi and H. Nakahashi, *J. Chromatogr.* **213,** 275 (1981).
240. W. Kullmann, *J. Liq. Chromatogr.* **2,** 1017 (1979).
241. T. F. Gabriel, J. Michalewsky, and J. Meienhofer, *J. Chromatogr.* **129,** 287 (1976).
242. T. F. Gabriel, M. H. Jimenez, A. M. Felix, J. Michalewsky, and J. Meienhofer, *Int. J. Pept. Protein Res.* **9,** 129 (1977).
243. F. Naider, R. Sipzner, A. S. Steinfeld, and J. M. Becker, *J. Chromatogr.* **176,** 264 (1979).
244. M. Goodman, P. Keogh, and H. Anderson, *Bioorg. Chem.* **6,** 239 (1977).
245. M. Furukawa, Y. Mori, Y. Takeuchi, and K. Ito, *J. Chromatogr.* **136,** 428 (1977).
246. A. Guyon, B. P. Roques, F. Guyon, A. Foucault, R. Perdrisot, J. Swerts, and J. Schwartz, *Life Sci.* **25,** 1605 (1979).
247. S. Ohlson, L. Hansson, P. O. Larsson, and K. Mosback, *FEBS Lett.* **93,** 5, (1978).
248. F. W. Frei and A. H. M. T. Scholten, *J. Chromatogr. Sci.* **17,** 152 (1979).
249. S. D. Lidofsky, T. Imasaka, and R. N. Zare, *Anal. Chem.* **51,** 1602 (1979).

Note Added in Proof

Macroreticular ion-exchange resins of the PS-DVB-type, having a relatively small particle size ($d_p \leq 6 \mu m$) and a high cross-linkage (>30%), have been employed for peptide mapping using gradients of increasing salt and organic salt concentration (*235a*). However, to ensure good peak efficiencies, temperatures up to 70°C are required. Recoveries of peptides in the range of 40–80% have been obtained with these resins. A new class of monodisperse ion exchangers for peptides have recently become available (*235b*). These ion exchangers are based on a hydrophilic, polyether resin with a narrow average particle diameter of the beads of 9.8 ± 2% μm, with a void volume of ca. 40% for packed columns. These new chromatographic media–known as Mono-Q, Mono-S, and Mono-P—exhibit very large pores with exclusion limits for globular proteins of approximately 10^7 daltons. Because of the porous nature of the matrix, high loading capacities are feasible without sacrificing resolution or separation speed. Several studies (*235b,c*) have reported their use for polypeptide separations with water–organic solvent combinations containing buffers and salts familiar to most biochemical workers. With the availability of these and similar high-capacity, chemically stable resins, the capability of chromatofocusing techniques has been considerably extended. In this procedure, peptides are eluted from an ion-exchange resin using a pH gradient internally generated with polymeric ampholytes such as the Ampholines or alternatively with multicomponent mixtures of simple buffers (*235d,e,f*). Polypeptides are separated on the basis of charge differences close to their isoelectric points with zone broadening minimized due to the intrinsic focusing effect of the internal pH gradient.

MOBILE-PHASE EFFECTS IN LIQUID–SOLID CHROMATOGRAPHY

Lloyd R. Snyder

Yorktown Heights, New York

I. INTRODUCTION

The variation of the mobile phase in liquid–solid chromatography (LSC)[1] for controlling the retention and separation of the sample has been

[1] By LSC we include separations on any polar, nonionic phase, e.g., alumina, silica, or bonded-phase packings such as cyanopropyl, aminopropyl, and diol-phase.

HIGH-PERFORMANCE LIQUID CHROMATOGRAPHY, Vol. 3

ISBN 0-12-312203-1

of interest to practical workers since the time of Tswett. The recognition of the importance of solvent polarity in determining solvent strength dates from the 1950s, and the value of equieluotropic series for maximizing separation selectivity was appreciated a few years later. Beginning in the 1960s, interest shifted to the development of more general models and quantitative theories of the retention process in LSC systems. The first comprehensive and detailed treatment of this kind—which I will refer to as the *displacement model*—appeared in 1968 (*1*). It was followed a few years later by the model of Soczewinski (*2*), and the models of Scott and Kucera (*3–6*) and of Jaroniec and co-workers (*7–10a*). The Soczewinski treatment represents a simplification of the original theory of Ref. *1*, and in fact is a special case of that more general theory (see Ref. *11*). The Jaroniec model, on the other hand, starts with Ref. *1* and incorporates additional detail—treating such effects as surface inhomogeneity of the adsorbent, interactions in the mobile phase, and other phenomena that are ignored in Ref. *1*. While some experimental verification of the Jaroniec model has been presented, it is not obvious that these additional effects actually play an important role in determining retention in typical LSC systems (most workers have so far focused on silica as adsorbent). The Scott–Kucera model differs fundamentally from the approaches followed by previous workers and those cited here. For a critical analysis of its assumptions and conclusions, see Ref. *12*.

The original displacement model of Ref. *1* has since been amplified and modified (*13–19*) in the light of new experimental data, and it has been used in general mobile-phase optimization strategies for maximizing sample resolution (*20, 21*). Similarly, the model of Soczewinski has been applied to numerous LSC systems (e.g., Refs. *22–27*), which has provided experimental verification of its ability to correlate and predict retention data for a wide range of sample types and mobile-phase compositions, by means of a simple equation. In the present article I will attempt to organize these recent developments into a single, comprehensive model for retention in LSC. This effort will be seen to provide new insights into the physicochemical processes that determine retention and its dependence on the composition of the mobile phase. As a result, certain puzzling observations on the role of the adsorbent in determining mobile-phase effects in LSC can now be reconciled with the known surface chemistry of the adsorbent, including alumina, silica, and organic-bonded-phase packings. It is also possible for the first time to explain these and other mobile-phase effects in terms of a single model that is derived more or less from first principles.

Previous treatments of the displacement model have for the most part ignored the effects of interactions between solute and solvent models in either phase. In this article we will reexamine the importance of these effects, including the case where hydrogen bonding between solvent and solute molecules is possible. We will also test the displacement model in two additional ways. First, the displacement model in its present form allows the prediction of isotherm data from retention data. That is, for a mobile phase A/B, where A and B are pure solvents, the concentration of solvent B in the stationary phase can be predicted as a function of its concentration in the mobile phase, on the basis of retention measurements for a solute as a function of the concentration of B in the mobile phase.

Second, previous tests of the displacement model have focused mainly on its ability to correlate and predict retention data in terms of derived correlational equations. Such correlations are based on various free energy relationships, and it is often found that comparisons of this kind can be insensitive to differences in the underlying physical model. That is, correlations of experimental retention data with theory may appear acceptable, in spite of marked deficiencies of the model. In some cases (e.g., Ref. *12,* sorption versus displacement models), radically different models can even yield the same or similar correlational equations. Here we will further test the proposed model for LSC retention in the following ways: (1) application of the model to a wide range of LSC systems, involving major variations in solute, solvent, and adsorbent; (2) examination of the various free energy terms that individually contribute to overall retention.

Finally, we will summarize here all the correlational equations and experimental solvent parameters required for predictions of solvent strength and selectivity in LSC, and discuss their significance in terms of mobile-phase optimization strategies.

II. THEORY

A. Displacement Model—Nonlocalizing Solutes and Solvents

This simple model of LSC retention has been discussed in detail (*1, 11, 12, 15*). It is reviewed here as a necessary beginning to the examination of more complex LSC systems, and to develop a consistent terminology for the general treatment of any LSC system.

Retention is assumed to occur as a displacement process, where an adsorbing solute molecule X displaces some number n of previously adsorbed mobile phase molecules M:

$$X_n + nM_a \rightleftharpoons X_a + nM_n \tag{1}$$

Subscripts "n" and "a" refer to a molecule in the nonsorbed and adsorbed phase, respectively. The equilibrium constant for reaction (1) is

$$K = (\mathrm{X})_\mathrm{a}(\mathrm{M})_\mathrm{n}^n/(\mathrm{X})_\mathrm{n}(\mathrm{M})_\mathrm{a}^n \tag{2}$$

The quantity $(i)_j$ is the mole fraction of species i in phase j. The dimensionless standard-state free energy of adsorption of X according to reaction (1) is

$$\Delta E_\mathrm{xm} = -\Delta G°/2.3RT = \log_{10} K \tag{3}$$

The quantity ΔE_xm can be related to the dimensionless partial molal free energies E_{ij} of each species i in each phase j [reaction (1)]:

$$\Delta E^*_\mathrm{xm} = E^*_\mathrm{xa} + nE_\mathrm{mn} - E_\mathrm{xn} - nE^*_\mathrm{ma} \tag{4}$$

Here, the superscript "*" refers to free energies for an adsorbent of standard activity; e.g., silica with some defined surface water content. Corresponding adsorption energies on some other surface of activity α' (relative to the standard adsorbent surface) are given as

$$\Delta E_\mathrm{xm} = \alpha'\Delta E^*_\mathrm{xm}, \qquad E_{i\mathrm{a}} = \alpha' E^*_{i\mathrm{a}} \tag{5}$$

Equations (5) follow from the functional relationship of interaction energies to properties of the two interacting species (*1*); α' is therefore a measure of the interaction strength of the adsorbent surface. It is believed that in most LSC systems the solution terms $E_{i\mathrm{n}}$ are effectively zero, as a result of cancellation of solvent–solute interactions in the two phases (*1*). Arguments in terms of solubility parameter theory lead to a similar conclusion (*28*).[2] Inserting Eqs. (5) into Eq. (4), with neglect of the solution terms, then gives

$$\Delta E_\mathrm{xm} = \alpha'[E^*_\mathrm{xa} - nE^*_\mathrm{ma}] \tag{6}$$

Values of the capacity factor k' are proportional to K,[3] so that k' values for a given solute in two different mobile phases 1 (k_1) and 2 (k_2) are given as

$$\log(k_1/k_2) = \alpha' n(E^*_\mathrm{m2} - E^*_\mathrm{m1}) \tag{7}$$

The quantity E^*_m2 refers to E^*_ma for mobile phase 2, and E^*_m1 is E^*_ma for mobile phase 1. If we define the cross-sectional area required on the adsorbent surface by the solute, mobile phase 1, and mobile phase 2 as A_s, A_1, and A_2, respectively, and if the mobile-phase adsorption energy per

[2] See the discussion of Section II,B,3.

[3] Here, $k' = (n_\mathrm{s}/n_\mathrm{m})K$, where n_s and n_m refer to the total moles of stationary phase (adsorbed monolayer) and mobile phase within the column, respectively.

unit area is ϵ°, equal to E^*_{ma} divided by A_m (equal to ϵ_1 or ϵ_2), then Eq. (7) yields

$$\log(k_1/k_2) = \alpha' A_s(\epsilon_2 - \epsilon_1) \quad (8)$$

The solvent strength ϵ° of mobile phase 1 is ϵ_1, and of mobile phase 2 is ϵ_2.

Equation (8) is a fundamental relationship for retention in LSC as a function of the solvent strength of the mobile phase. It states that log k' values for different solutes will yield linear plots against values of ϵ° for different mobile phases, and the slopes of these plots will be proportional to the molecular size A_s of the solute. Numerous data are summarized or referenced in Ref. *1*, showing the validity of Eq. (8) when applied to LSC systems where the solute and solvent molecules are nonlocalizing (nonpolar or moderately polar compounds—see Section II,B below). Similar data showing the applicability of Eq. (8) to amino-phase polar-bonded-phase columns are given in Ref. *17*.

The solvent strength ϵ° of a multisolvent mobile phase can be related to ϵ° values of the pure constituent solvents in the mixture. Reactions similar to reaction (1) can be written for the equilibrium between different solvent molecules A, B, C, . . . (assuming $\epsilon_A > \epsilon_B > \epsilon_C$ ...), assuming n is equal to one for all solvent pairs:

$$B_n + A_n \rightleftharpoons B_a + A_n \quad (9a)$$

$$C_n + B_a \rightleftharpoons C_a + B_n \quad (9b)$$

$$\vdots$$

Equilibrium constants such as those in Eq. (2) can then be derived for reactions (9a), (9b), . . . :

$$K_{ba} = (B)_a(A)_n/(B)_n(A)_a$$

$$= 10^{\alpha' n_b(\epsilon_B - \epsilon_A)} \quad (10a)$$

$$K_{cb} = 10^{\alpha' n_b(\epsilon_C - \epsilon_B)} \quad (10b)$$

$$\vdots$$

If the surface mole fraction θ_i of solvent i is defined, the sum of all θ_i values (for the various solvents in the mobile phase mixture) must be equal to one. This then allows the calculation of ϵ° for the mobile-phase mixture A/B/C ... as (*15*)

$$\epsilon^\circ = \epsilon_A + \log(N_A/\theta_A)/\alpha' n_b \quad (11)$$

Here, the mole fraction of i in the mobile phase is given as N_i, and n_b is the molecular cross-sectional area (A_s value) of solvent molecules. When the various solvents are of different size, n_b is the average value of A_m for each solvent (B, C, ...) exclusive of nonpolar solvents (e.g., A).[4] Equation (11) is general for any number of solvents in the mobile phase (*15*). For the special case of a binary-solvent mobile phase A/B, Eq. (11) can be expressed as

$$\epsilon^\circ = \epsilon_A + \frac{\log(N_b 10^{\alpha' n_b(\epsilon_B - \epsilon_A)} + 1 - N_B)}{\alpha' n_b} \tag{11a}$$

Equation (11) has been verified for numerous mobile phases and alumina, silica, or amino-bonded phase as adsorbent, using nonlocalizing solvents A and B (*1, 14–17*).

The application of Eq. (11) to an experimental set of ϵ° values for a multisolvent mobile phase also allows the calculation of θ_i for each solvent as a function of mobile-phase composition (N_A, N_B, N_C, ...). (*15*). Thus, the present model allows the calculation of experimental isotherms for binary-solvent or more complex mobile-phase systems and a given adsorbent. Values of ϵ_B (or other ϵ_i value) can also be determined as a function of θ_B, to test the assumption that ϵ_B is constant as θ_B is varied (which simple theory predicts).

The model just described is further illustrated in Fig. 1a, for adsorption of nonlocalizing molecules of solute X or mobile phase M within the adsorbed monolayer. Discrete adsorption sites are shown (asterisks in Fig. 1), but adsorbed molecules are assumed not to prefer positions over these sites; i.e., the adsorbent surface can be considered as homogeneous. This model appears to work well for less polar solvents and solutes, but for many LSC systems it begins to fail as adsorbate polarity increases.

[4] For consistency, the value of n_b assumed should be given by

$$n_b = \frac{\theta_B n_B + \theta_C n_C + \cdots}{(1 - \theta_A)} \tag{11b}$$

where n_B, n_C, ... are the A_m values of solvents B, C, ... Equation (11b) weights the value of n_b according to the relative adsorption of different polar solvents B, C, ... For binary-solvent mobile phases A/B, $n_b = n_B$. For multisolvent mobile phases A/B/C ... , Eq. (116) is intuitively reasonable, it avoids problems with discontinuous variations in n_b as mobile-phase composition is varied from ternary to binary compositions, and it is somewhat more accurate for predicting solute retention data versus a simple average of n_B, n_C, ... (*15, 16*). It should be noted that Eq. (11) for multisolvent mobile phases *assumes* that each solvent in the mixture has the same molecular cross section (value of A_m or n_b), which is seldom the case. However, Eq. (11) is a rather good approximation to actual LSC mobile phases. Concerning the effect of variation in A_s, n_b, or n_i on the adsorption equilibrium, see also the discussion of Jaroniec and Oscik (*28a*).

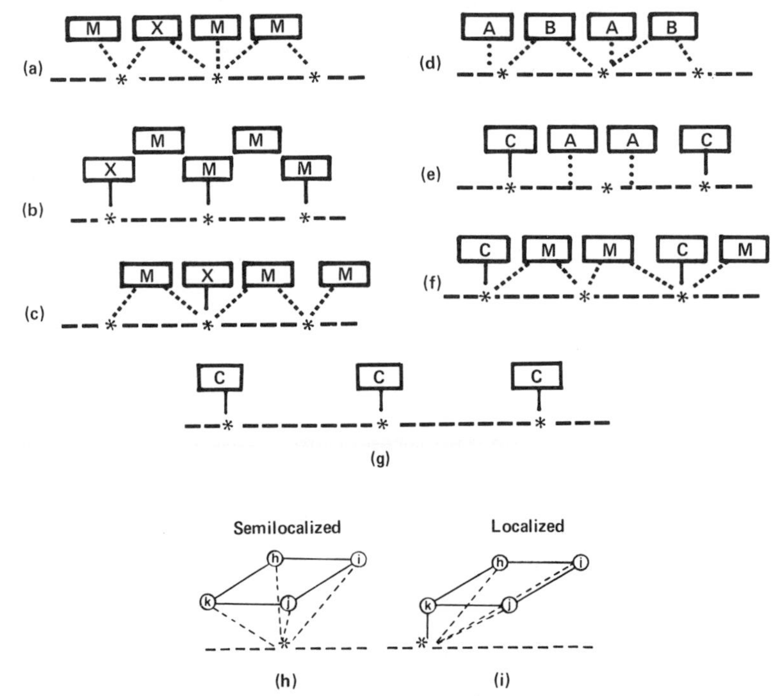

FIG 1. Representation of adsorbed molecules of solute (X), mobile phase (M), and solvents A (nonpolar), B (polar, nonlocalizing), and C (polar, localizing): (a) nonlocalized molecules X and M; (b) localizing molecules X and M (some M molecules are delocalized); (c) localized X, nonlocalized M; (d) nonlocalized A and B; (e) localized C, nonlocalized A; (f) localized C, nonlocalized M; (g) localized C, smaller concentration of surface sites; (h) nonlocalized solute h–i–j–k; (i) localized solute h–i–j–k; (—) localizing (polar) interaction, (---) nonlocalizing (polar) interaction, (···) nonpolar (weak) interaction.

Reasons for this failure are further discussed in the following sections (II,B; II,C), and can be related to the following physical phenomena:

(1) The localization of polar molecules over discrete adsorption sites—illustrated for the lower three molecules of mobile phase M in Fig. 1b, and for the solute X in Fig. 1c.
(2) Interactions among solvent and solute molecules within the adsorbed monolayer, apart from interactions that are canceled by similar interactions in the mobile phase.
(3) Interactions among solvent and solute molecules in the mobile phase, apart from such interactions that are canceled by corresponding adsorbed-phase interactions.

B. Displacement Model Including Localization

As the polarity of adsorbing molecules of solvent or solute increases, interactions of these molecules with the polar adsorbent surface become stronger, and there is an increasing tendency for the molecule to adsorb with localization (*1*, *12*), i.e., with the adsorbate molecule centered over a site, as just discussed for X and M in Fig. 1b and c. Figure 2 illustrates this in further detail, for the specific case of adsorption of a localizing solvent molecule C onto silica, from pure mobile phase C. Surface silanols constitute the adsorption sites on silica, as shown in Fig. 2a. These are randomly arranged on the amorphous silica surface, as shown in the overhead view of Fig. 2b, where individual sites are indicated by asterisks. Actual solvent molecules C are not mathematical points in space, but comprise a finite volume that is apportioned between polar and nonpolar parts. A simple case is that of a compound C which possesses a single polar group k, as illustrated in the side view of Fig. 2c. The group k might be, for example, the —O— group in diethyl ether. In Fig. 2c the group k is arbitrarily assumed to be centered on the face of a disc-shaped molecule of C, and localization then corresponds to alignment of the group k over an adsorption site. In Fig. 2c, localized adsorption of C is shown for molecules *i*, but the in-between molecule *ii* is not localized, because the surrounding adsorbed molecules prevent the alignment of the group k (in the in-between molecule) directly over an adsorption site. Because this vertical positioning of localizing groups k and an adsorption site[4a] is shifted away from the optimal (aligned) configuration, the energy of adsorption of the adsorbate molecule is correspondingly decreased. That is, the energy of adsorption of localized molecules is usually much larger than for the same nonlocalized molecules. This situation is similar to the energy changes that accompany the distortion or lengthening of a covalent bond between two atoms.

1. Restricted-Access Delocalization of the Solvent

There are several processes that can lead to the delocalization of an adsorbate molecule that prefers to adsorb with localization. One such process is illustrated in Fig. 2c, for the in-between molecule *ii* discussed above. This behavior of the molecule *ii* is referred to as *restricted-access delocalization* of the solvent molecule C (or M). Figure 2c is shown as an overhead view in Fig. 2d. Delocalized molecules C (shown as dashed circles) arise either from steric hindrance by surrounding localized molecules (*ii*b) or from the inability of C to center itself over a site as in *ii*a and Fig. 2c.

[4a] See discussion of Fig. 10-3, Ref. (*1*).

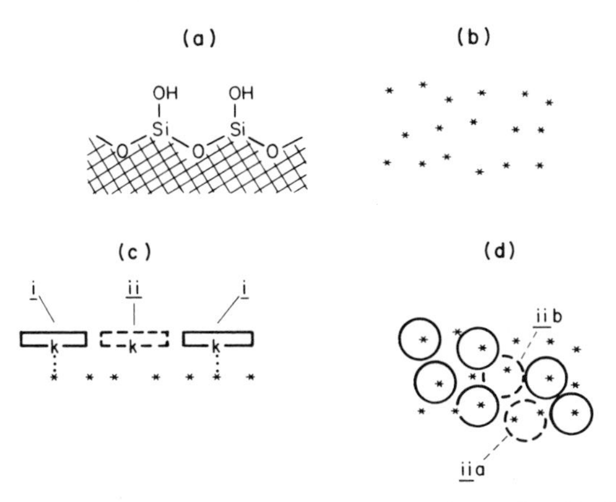

FIG. 2. Representation of adsorption of localizing solvent C: (A) silica surface; (B) overhead view of silanols on silica surface; (C) side view of adsorbed molecules C with localizing group k, where $k \equiv$ —O—, —CHO, —CO_2CH_3, etc; ii, delocalized; (D) overhead view of (C). Reprinted with permission from Snyder *et al.* (*18*).

For surfaces such as alumina or silica, which are rigid and covered with a high density of adsorption sites, it appears that about 75% of the molecules C in the adsorbed monolayer can adsorb with localization; the remaining molecules of C suffer restricted-access delocalization. This can be inferred from three different kinds of experimental observation, involving mobile-phase mixtures A/C (*14, 15, 18*). First, isotherm measurements (such as those in Refs. *4* and *5*; see also discussion of Ref. *12*) suggest that the adsorption of C occurs in two steps; about 75% of a monolayer is adsorbed initially (localized), whereas the adsorption of the last 25% of C (delocalized) occurs with a distinctly lower energy of adsorption of C. Second, experimental values of $\epsilon°$ can be determined for various values of N_A [via Eq. (8)], and values of ϵ_C ($\equiv \epsilon_B$) can in turn be determined from Eq. (11a) as a function of θ_C. For localizing solvents C, in mixtures A/C, it is observed that the value of ϵ_C is higher at low values of θ_C (localization of C) than for $\theta_C \approx 1.0$ (delocalization of C), and the change in $\epsilon_C (d\epsilon_C/d\theta_C)$ is greatest in the region $0.7 > \theta_C > 0.85$, where completion of a localized monolayer is taking place. Finally, solvent selectivity has been related (*13, 18, 19*) to competition of localizing solvent and solute molecules for adsorption sites. Theory predicts that such changes in solvent selectivity (value of m—see below) should correlate with the number of adsorption sites on which solute localization can occur, without having to displace a molecule of localized C. It is found that m increases with θ_C, up to a maximum value at about $\theta_C = 0.75$, which again implies completion of the layer of localized molecules C for $\theta_C = 0.75$.

This theoretical model for restricted-access delocalization further shows that values of ϵ_C are predictable as a function $\%_{lc}$ of θ_C,

$$\epsilon_C = \epsilon''_C + \%_{lc}(\epsilon'_C - \epsilon''_C) \tag{12}$$

Here, ϵ'_C is the value of ϵ_C for localized C at low values of θ_C, and ϵ''_C is the value for delocalized C at high values of θ_C. The ratio ϵ'_C/ϵ''_C should equal one for less polar, nonlocalizing solvents, and this is observed. Likewise, this ratio should increase for increasing localization of C (or increasing adsorption energy for the group *k* in the molecule C), and this too is observed (*14*). Finally, Eq. (12) allows the calculation of ϵ_C as a function of θ_C, which is required in the calculation of $\epsilon°$ for multisolvent mobile phases that contain localizing solvents C.

There are two requirements that the adsorbent surface must meet for restricted-access delocalization to occur. *First,* there must be a higher surface density of adsorption sites than of localizing solvent molecules that can be fit into the adsorbed monolayer (space-filling considerations). This is illustrated in Fig. 1g, for a surface with a rather low concentration of surface sites. Here it is assumed that there is no specific interaction possible between an adsorbing molecule and the surface, except through the adsorption site. *Second,* the adsorption sites must be rigidly positioned on the surface. If they are free to vary their position [for example, in polar-bonded-phase packings where the group (*) is attached to an alkyl chain], then restricted-access delocalization will occur to a lesser extent than for rigidly positioned sites.

2. *Site-Competition Delocalization of the Solute*

Less-polar solvent molecules B ($CHCl_3$, CH_2Cl_2, benzene, etc.) that do not localize nevertheless interact with adsorption sites. This is illustrated in Fig. 1d for the binary mixtures A/B (A is nonpolar), and is contrasted in Fig. 1e for adsorption of a mobile phase A/C, where C is localizing. When nonlocalizing molecules of a polar mobile phase M are adjacent to localizing molecules of solute X (Fig. 1c) or solvent C (Fig. 1f), these noncovalent interactions of M with surface sites can interfere with or displace corresponding interactions between the localized molecule and its site. This effect is referred to as *site-competition delocalization.*

Alternatively, it can be argued that these lateral (noncovalent) interactions of a nonlocalizing solvent M with the surface are simply preempted by a localizing molecule C or X. This then represents a loss in energy of the final system (adsorption of C or X) which is equivalent to that proposed above and in Figs. 1c,f. See the further discussion of Ref. (28b).

Site-competition delocalization manifests itself in two ways: (a) an increase in the apparent solute molecular size, or value of A_s in Eq. (8) and

(b) a dependence of the value of ϵ_C' (localized-solvent adsorption energy per unit area—at small values of θ_C) on the solvent strength ϵ_B of the nonlocalizing solvent B in mixtures B/C. Consider first the effect on the apparent A_s value of the solute.

Assume a mobile phase M which is a solvent mixture A/B; A is nonpolar ($\epsilon_A = 0.0$) and B is polar ($\epsilon_B > 0.0$) but nonlocalizing. Let the solute adsorption energy E_{xa} be given as E_{xa}° for the case where $\epsilon_M = 0.0$ (pure A as mobile phase). Now consider how E_{xa} changes with change in the solvent strength ϵ_M of the mobile phase M. From Fig. 1c it is seen that both mobile-phase molecules M and localizing-solute molecules X compete for the same adsorption site. X competes by direct (covalent) interaction, whereas M competes indirectly by lateral interactions as shown. It is intuitively reasonable that the interaction of X with a site is decreased in proportion to both the strength of interactions between M and the site, and the relative localization of X [when X is nonlocalizing, E_{xa} is independent of ϵ_M—see Eq. (8)]. The strength of interactions between all surrounding molecules M and the site occupied by localized X will be proportional to the adsorption energy of M per unit area, or ϵ_M. Similarly, the relative localization of X can be described by a localization function $f_1(X)$, which should increase with E_{xa}° or the adsorption energy Q_k° of a localizing group k within the molecule X. Thus,

$$E_{xa} = E_{xa}^\circ - f_1(X)\epsilon_M \tag{13}$$

Equation (6) can be combined with Eq. (13) to give[5]

$$E_{xm} = \alpha'[E_{xa}^\circ - f_l(X)\epsilon_M - nE_{ma}] \tag{13a}$$

which, as in the derivation of Eq. (8), can be written

$$\begin{aligned} E_{xm} &= \alpha'[E_{xa}^\circ - f_1(X)\epsilon_M - A_s\epsilon_M] \\ &= \alpha'\{E_{xa}^\circ - [A_s + f_l(X)]\epsilon_M\} \end{aligned} \tag{13b}$$

Values of k' are proportional to K, so that $\log(k_1/k_2)$ for mobile phases 1 and 2 is given as

$$\log(k_1/k_2) = E_{x1} - E_{x2}$$

where E_{x1} and E_{x2} refer to values of E_{xm} for mobile phases 1 and 2, and therefore [from Eq. (13b)]

$$\begin{aligned} \log(k_1/k_2) &= \alpha'[A_s + f_l(X)](\epsilon_2 - \epsilon_1) \\ &= \alpha'(A_s)_{expt}(\epsilon_2 - \epsilon_1) \end{aligned} \tag{14}$$

[5] For simplicity, here and later we have dropped the "*" from the various terms E^*. The differentiation of standard-state from less-active adsorbents will generally be clear from the context.

Comparing Eq. (14) with Eq. (8), we see that the two equations are identical, if we replace A_s in Eq. (8) by $[A_s + f_1(X)] \equiv (A_s)_{expt}$.[6] That is, when localizing solutes X are subjected to site-competition delocalization, the *apparent* value of $A_s = (A_s)_{expt}$ in Eq. (8) is larger than estimated from the molecular dimensions of X.

Site-competition delocalization of solutes X has been observed for adsorption of localizing solutes X onto silica and amino-bonded-phase packings, but not for alumina (*1, 17*). We will comment on this later.

The requirement for site-competition delocalization of the solute would appear to be an adsorption site which allows both (a) the localization of a solute molecule X, and (b) the lateral interaction with the site by an adjacent mobile phase molecule M—as in Fig. 1c. Thus, the presence of this phenomenon for some LSC systems (e.g., silica) and not others (e.g., alumina) implies some fundamental difference in the relative accessibility of the adsorbent sites to both solvent and solute molecules.

3. *Site-Competition Delocalization of the Solvent*

When a binary-solvent mobile phase B/C is used, where C is localizing and B is not, the value of E°_{ca} for the localized solvent C [analogous to E°_{xa} for a localized solute X, as in Eq. (13)] will vary with ϵ_B, just as in the case of site-competition delocalization of the solute. This has been discussed in Ref. *16*, and the resulting relationship is [cf. Eq. (13)][7]

$$\epsilon'_C = (\epsilon'_C)^{\circ} - [f_l(C)/A_c]\epsilon_B \qquad (15)$$

Here, $(\epsilon'_C)^{\circ}$ refers to the value of ϵ'_C for $\epsilon_B = 0.0$, $f_l(C)$ is the same function as $f_l(X)$, only its value is determined by the relative localization of C, rather than X, and A_c is the value of A_m for the molecule C. Equation (15) has been verified for a number of localizing solvents C and silica as adsorbent. Thus, values of ϵ'_C have been found to decrease with increase in ϵ_B, for mobile phase B/C, and the function $f_l(C)$ has been found to be identical to the function $f_l(X)$, depending in each case on the value of E°_{xa} or E°_{ca} (*16*). Knowing the function $f_l(i)$ (*16*, Fig. 3), where i is either a solvent or solute molecule, it is possible to calculate values of ϵ'_C (for mixtures B/C) as a function of ϵ_B, or values of $(A_s)_{expt}$ as a function of the solute X.

4. *Intramolecular Delocalization of Solute or Solvent*

Still another manifestation of the general phenomenon of localization and delocalization is encountered in molecules of solute or solvent which

[6] In Ref. *1*, $f_l(X)$ is referred to as Δa_i for monofunctional solutes.

[7] Equation (15) results from Eq. (13) by substituting E_{ca} and E°_{ca} for E_{xa} and E°_{xa}, respectively, then dividing both sides of the equation by A_c.

possess more than one adsorbing functional group *i*, as well as at least one localizing group *k*. This is illustrated for the model adsorbate *hijk* shown in Fig. 1h and i. When all adsorbate groups are nonlocalizing, as in Fig. 1h, the molecule as a whole is nonlocalized. When one (or more) solute group *k* in the molecule is sufficiently polar to localize, localization of *k* occurs as in Fig. 1i, with concomitant intramolecular delocalization of remaining adsorbate groups *h*, *i*, and *j*.[8] Such delocalization effects result in a failure of the Martin equation (*29*),

$$\log k' = R_M = \sum \Delta R_M \tag{16}$$

as discussed in Refs. *1* and *30*. Thus, the group-retention factors ΔR_M for different solute groups *i* as measured in monofunctional molecules X do not sum to give the value of $\log k'$ predicted by Eq. (16), when one (or more) group *i* is capable of localization.

Intramolecular delocalization should be more pronounced for adsorbents with a low density of surface sites, and less important for adsorbents with more exposed and/or more flexibly positioned sites.

C. Interactions among Solvent and Solute Molecules—Effect on $\epsilon°$

The present model has so far assumed that interactions between molecules of solvent or solute in either phase can be ignored. Now we will examine the effects of these interactions on retention in various LSC systems. Equation (4) for the retention of a solute X in a mobile phase M recognizes intermolecular interactions in the mobile phase (n), but assumes that adsorbed-phase free energies (E_{ia}) are not a function of intermolecular interactions within the adsorbed phase.[9] We can recognize these adsorbed-phase intermolecular interactions by adding an energy term $E_{ia}{}^{m}$ to Eq. (4) for each adsorbed species *i*:

$$\Delta E_{xm} = E_{xa} + E_{xa}{}^{m} + nE_{mn} - E_{xn} - n(E_{ma} + E_{ma}{}^{m}) \tag{17}$$

We can express the above energy terms ($E_{ia}{}^{m}$, E_{in}) due to interactions among solvent and solute molecules in terms of activity coefficients γ_{ij} for each compound *i* in each phase *j*:[10]

[8] When groups *h*, *i*, and *j* are not sufficiently polar to localize, it might be assumed that they are also free from intramolecular delocalization. Experimentally this is not the case (*1*), so it must be assumed that the constraint on these groups which is imposed by localization of *k* nevertheless reduces the adsorption energy of less polar groups *h*, *i*, and *j*.

[9] Note that E_{ia} can change with mobile-phase composition due to site-competition delocalization, but this effect does not arise from solvent–solute or solvent–solvent interactions. Its effect on retention was discussed in Sections II,B,2 and 3.

[10] γ_{ij} is defined versus pure hexane mobile phase as standard state for each phase.

$$\log \gamma_{xn} = -E_{xn} \tag{17a}$$

$$\log \gamma_{mn} = -E_{mn} \tag{17b}$$

$$\log \gamma_{xa} = -E_{xa}{}^{m} \tag{17c}$$

$$\log \gamma_{ma} = -E_{ma}{}^{m} \tag{17d}$$

Equation (6), which is the basis of the retention model so far discussed, can now be corrected for intermolecular interactions by combining Eqs. (5) and (17):

$$\Delta E_{xm} = \alpha' E_{xa} - \alpha' n E_{ma} - \log Q \tag{18}$$

where

$$Q = (\gamma_{xa}/\gamma_{xn})(\gamma_{mn}/\gamma_{ma})^{n} \tag{18a}$$

The effect of the log Q term on LSC retention can be expressed as an equivalent change in the solvent strength of the mobile phase, from the value $\epsilon^\circ = E_{ma}/A_m$ defined previously [Eqs. (7) and (8)] to an apparent value ϵ^m given by[11]

$$\epsilon^m = E_{ma}/A_m + (\log Q)/\alpha' A_s \tag{19}$$

When Eq. (19) is substituted into Eq. (18), we obtain

$$\Delta E_{xm} = \alpha' E_{xa} - \alpha' A_s \epsilon^m$$

which then yields equations that are equivalent to Eq. (8):

$$\log(k_1/k_2) = \alpha' A_s(\epsilon_2 - \epsilon_1) + \log(Q_2/Q_1) \tag{19a}$$

or

$$\log(k_1/k_2) = \alpha' A_s(\epsilon_2{}^m - \epsilon_1{}^m) \tag{19b}$$

Experimental values of ϵ° for different solvents have in the past been determined by applying Eq. (8) [or Eq. (19b)] to specific data. Therefore, these values correspond to values of ϵ^m. Elsewhere in this article we will not differentiate between values of ϵ^m and ϵ°, arguing rather that the difference between these two quantities is normally small (i.e., $Q \approx 1.0$). However, in Section II,A,4, where we return to a discussion of the magnitude of Q and its effect on ϵ° values, we will examine the approximation $\epsilon^\circ \approx \epsilon^m$ in terms of the constancy of values of ϵ_B in mixtures A/B.

Various workers have considered the effect of the log Q term on retention in LSC systems (usually for silica) (*2, 8, 31, 32*). Soczewinski (*2*)

[11] Of course, ϵ^m values will be solute-specific, due to the inclusion of solute-specific terms in log Q. The simple displacement model predicts that solvent ϵ° values are not a function of the solute.

recognized the possible importance of solvent–solute interactions in the mobile phase (γ_{mn}), and proposed the estimation of its effect on retention by assuming that X–M complexes are formed in the mobile phase. Hennion *et al.* (*31*) give a similar treatment for amino-phase adsorbents. In each case, the correction for such solvent–solute complexation assumes a fairly simple mathematical form. Poppe *et al.* (*32*) considered all interactions in the mobile phase (γ_{xn}, γ_{mn}) and measured actual values of γ_{in} for direct correction for these effects in the displacement model. Finally, Jaroniec *et al.* (*8*) considered all four γ_{ij} values and estimated their contribution to LSC retention by assuming various complexation reactions: $M + M \rightleftharpoons M_2$ or $X + M \rightleftharpoons XM$, in each phase. The resulting theoretical expressions are complex, and their practical application to actual LSC systems (where all γ_{ij} terms may be significant) is not obvious.

In Section III,A,4 we will examine the actual importance of the Q term of Eq. (19) as it affects experimental values of $\epsilon°$. On theoretical grounds, however, we can note certain reasons for expecting that this term might be small in typical LSC systems. For simplicity, assume a pure solvent M as mobile phase,[12] and solute and solvent molecules of equal size ($n = 1$).

First, note that adsorbed molecules X or M are surrounded by molecules M within the adsorbed monolayer (e.g., Fig. 1a), as well as by molecules M in the mobile phase adjacent to the adsorbed monolayer. Thus, the possible interactions among these molecules to a considerable extent cancel similar interactions in the mobile phase, so that values of $(\gamma_{xn}/\gamma_{xa})$ and $(\gamma_{mn}/\gamma_{ma})$ are much smaller than are values of γ_{xn} or γ_{mn}.

Second, if the two compounds X and M are not too different in polarity (which is often the case in LSC systems), then theoretical arguments based on solubility parameter theory (e.g., Ref. *28*) suggest that $\gamma_{xn} \approx \gamma_{mn}$ and $\gamma_{xa} \approx \gamma_{ma}$. Each of these two effects has the result of decreasing the value of Q [Eq. (18a)] relative to individual values of γ_{ij}. However, when interactions among molecules X and M involve hydrogen bonding, these arguments for the unimportance of Q must be reexamined (see the following section).

Hydrogen Bonding between Solute and Solvent Molecules

There is considerable evidence to suggest that solute–solvent interactions can no longer be neglected when hydrogen bonding between solute and solvent molecules is possible (*1, 11, 13, 19*). For the case of a proton-donor solute X–H and a basic solvent molecule C:, formation of

[12] Similar arguments can be applied for the mixed mobile phase A/B. However, it must be emphasized that a good understanding of why solute–solvent and solvent–solvent interactions effectively cancel in many LSC systems has not yet been fully achieved.

hydrogen-bonding complexes is possible in both the mobile and adsorbed phases:

$$\text{(mobile phase)} \qquad XH_n + C_n \rightleftharpoons XHC_n \qquad (20a)$$

$$\text{(adsorbed phase)} \qquad XH_a + C_a \rightleftharpoons XHC_a \qquad (20b)$$

In the adsorbed phase, molecules of XH or C can be either localized or delocalized, which further complicates a general description of the effects of hydrogen bonding on retention. However, compounds capable of strong hydrogen-bonding interactions will generally be relatively polar, so a reasonable first approximation is that hydrogen bonding in the adsorbed phase involves localized molecules, when $\theta_C \leqslant 0.75$. For the case of solute XH and solvent C: (see Fig. 1), we can then portray this situation in the adsorbed phase as

$$\underset{*}{\underset{|}{XH}} + \underset{*}{\underset{|}{C:}} \rightleftharpoons \underset{*}{\underset{|}{XH}}\cdots\underset{*}{\underset{|}{:C}} \qquad (21)$$

Because of steric constraints on the hydrogen bonding of localized molecules XH and C:, as well as possible involvement of the electron pair on C: with the adsorbent site (*), the equilibrium constant for reaction (20b) should be smaller than for reaction (20a).

The equilibrium constants for reactions (20a) and (20b) can be written as

$$\text{(mobile phase)} \qquad K^{cx} = (XHC)_n/(XH)_n N_c \qquad (22a)$$

and

$$\text{(adsorbed phase)} \qquad K^{cx*} = (XHC)_a/(XH)_a \theta_C \qquad (22b)$$

Larger values of K^{cx} and K^{cx*} mean stronger hydrogen bonding of solute and solvent molecules in the mobile and adsorbed phases, respectively. Additionally, we can write for adsorption of XH from the mobile phase (assume $A_s = n_b$ for convenience):

$$XH_n + C_a \rightleftharpoons XH_a + C_n$$

$$K_{xc} = [(XH)_a/(XH)_n]_c \qquad (23)$$

where K_{xc} is a function of the LSC system, i.e., solvent strength $\epsilon°$. The capacity factor k_{xc} for adsorption of XH from the mobile phase A/C can be expressed as

$$k_{xc} = (V_a/V_n)\left[\frac{(XH)_a + (XHC)_a}{(XH)_n + (XHC)_n}\right]_c \qquad (24)$$

Combining Eqs. (22a), (22b), (23), and (24) then yields

$$k_{xc} = (V_a/V_n)K_{xc}\left[\frac{1 + K^{cx*}\theta_C}{1 + K^{cx}N_B}\right] \tag{25}$$

Here V_a and V_n refer to the volumes of stationary and mobile phases, respectively. This model of solute–solvent hydrogen bonding in LSC is pursued further in Section III,C.

D. Solvent Selectivity Effects

Solvent selectivity refers to the ability of different mobile phases to change the relative retention or separation factor α of two solutes X and Y. Thus, if we define capacity factors k_{xb}, k_{xc}, k_{yb}, and k_{yc} for solutes X and Y in the mobile phases A/B and A/C, respectively, and if separation factors α_b and α_c are defined as

$$\alpha_b = k_{xb}/k_{yb} \tag{26a}$$

$$\alpha_c = k_{xc}/k_{yc} \tag{26b}$$

then differing solvent selectivities for mobile phases A/B and A/C will result in differing values of α_b and α_c. Thus, if two solutes are not separated by one mobile phase ($\alpha = 1$), a mobile phase of differing selectivity will effect the separation of the two compounds. We will next examine various contributions to solvent selectivity.

1. Non-Hydrogen-Bonding Systems

A change in solvent strength for a given mobile phase A/B (by varying the concentration of B) will be accompanied by a change in solvent selectivity whenever the A_s values of X (A_x) and Y (A_y) are different. This can be seen in terms of Eq. (8) (with appropriate changes in notation):

$$\log(k_{xc}/k_{xb}) = \alpha' A_x(\epsilon_b - \epsilon_c) \qquad \text{(for X)} \tag{27a}$$

$$\log(k_{yc}/k_{yb}) = \alpha' A_y(\epsilon_b - \epsilon_c) \qquad \text{(for Y)} \tag{27b}$$

for mobile phases A/B and A/C. Defining separation factors for X and Y in mobile phases A/B and A/C as in Eqs. (26a) and (26b), we have from Eqs. (27a) and (27b)

$$\alpha_c/\alpha_b = \alpha'(A_x - A_y)(\epsilon_b - \epsilon_c) \tag{28}$$

Thus, for different solvent strengths ϵ_c and ϵ_b, and different A_s values A_x and A_y, $\alpha_c/\alpha_b \neq 1$; i.e., a change in α results between mobile phases A/B and A/C. This change in solvent selectivity will be greater, the greater the difference in the A_s values of the two solutes and the greater the difference in their solvent strengths.

Generally we want to optimize the value of ϵ° for a mobile phase in the case of a given sample, so as to provide sample k' values in the optimum range of roughly 1–10 (e.g., *20*). This means that we are normally limited in how much we can vary ϵ° to achieve useful changes in selectivity.[13] Furthermore, the more difficulty separable solute pairs will generally be similar in their chemical and physical properties, which implies similar A_s values. Consequently, *simple variation in solvent strength is quite limited in its ability to achieve useful differences in solvent selectivity and resulting changes in* α *for hard-to-separate solute pairs.*[14]

Much larger changes in solvent selectivity can be effected by so-called *solvent–solute localization* (*18*). The latter phenomenon arises from differences in the relative localization of solute and solvent molecules in various LSC systems, as illustrated in Fig. 3. Here we assume a moderately polar (nonlocalizing) solvent molecule B and a polar (localizing) solvent molecule C. The solute X is localizing and the solute Y is nonlocalizing. The resulting arrangement of adsorbed solvent molecules within the adsorbed monolayer is shown for A/B in Fig. 3a and for A/C in Fig. 3b. Localized molecules are shown as solid circles (C), whereas nonlocalized molecules are shown as broken triangles (A) or squares (B). The adsorption of a solute molecule X (localizing) and Y (nonlocalizing) from the mobile phase A/B is illustrated in Fig. 3c,d. In either case, because A and B are nonlocalizing, the adsorbing solute molecule displaces a nonlocalized molecule of A or B. In the case of the adsorption of X or Y from a mobile phase A/C that is localized (Fig. 3e,f), X adsorbs with localization and must therefore displace a preadsorbed molecule of localized C. Because Y is nonlocalizing, it adsorbs by displacing a nonlocalized molecule of A or C. If we assume that X and Y have the same k' values ($\alpha_b = 1$) in the system A/B (Fig. 3c,d), then the k' value of X must be less than that of Y ($\alpha_c \neq 1$) in the system A/C (Fig. 3e,f). The reason is that the free energy required to displace a solvent molecule in Fig. 3c,d is the same for both X and Y, because for each solute the displaced solvent molecule (A or B) is delocalized. In Fig. 3e,f, however, solute X (but not Y) must displace a localized molecule of C during adsorption, and the energy required for this will be greater than for displacement of a delocalized molecule of C (or A) by an adsorbing molecule of Y.

[13] This limitation could be overcome by varying adsorbent surface area so as to change solute k values independently of ϵ°. However, this is a more complex selectivity-optimization strategy.

[14] Perry (*33*) has argued to the contrary, suggesting that change in ϵ° is in fact an effective means of changing α. See Ref. *33* for the pros and cons of this issue.

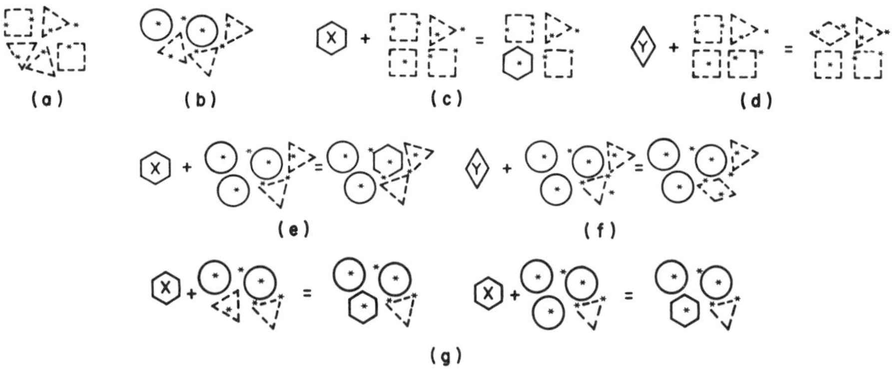

(a) (b) (c) (d)

(e) (f)

(g)

FIG. 3. Solvent–solute localization and separation selectivity. (a) representation as in Fig. 2d of adsorption of mobile phase A/B (△, —nonpolar; □, —polar, nonlocalizing); (b) same for mobile phase A/C (C—polar, localizing); (c) adsorption of localizing solute X from mobile phase A/B; (d) adsorption of nonlocalizing solute Y from mobile phase A/B; (e) adsorption of X from mobile phase A/C; (f) adsorption of Y from mobile phase A/C; (g) completion of localized layer of molecules C at $\theta_c = 0.75$ with maximum effect on solvent–solute localization selectivity.

As a result of solvent–solute localization, *a change from a localizing mobile phase A/C to a nonlocalizing mobile phase A/B can create large differences in solvent selectivity,* and the α values of various solute pairs. The effect is limited to solutes which show some degree of localization, and is therefore more pronounced for more polar samples and the stronger mobile phases that are required for their optimum separation.

We can derive a quantitative model of solvent–solute localization as follows. In the general case, both solutes and solvents will exhibit varying tendencies toward localization, rather than being characterizable as simply "localizing" or "nonlocalizing." Thus, the selectivity effects of solvent–solute localization will increase with increasing tendencies toward localization of solute X and solvent M (i.e., increase in polarity of the localizing group k in molecules X or M). The effect of solvent localization will also be more pronounced for higher mole fractions N_C of the localizing solvent in the mobile phase A/C, as illustrated in Fig. 3g. Here, for a lower value of N_C and a resulting value of $\theta_C = 0.5$, a molecule of X can adsorb with localization by displacing a delocalized molecule of A. However, for a higher value of N_C, such that $\theta_C = 0.75$, localized adsorption of X requires displacement of a localized molecule of C. The reason is that with increase in θ_C from 0 to 0.75, all adsorbing molecules C (or X) can localize onto the surface. When θ_C equals a value of 0.75, additional molecules of C adsorb without localization, and molecules of a localizing solute X must then displace molecules of localized C. Therefore, the effect of solvent localization on selectivity will increase in magnitude with increase in N_C and θ_C, until $\theta_C \sim 0.75$.

Equation (8) [or Eq. (14)] already recognizes the localization of solvent and solute. Thus, localization of the solvent leads to a predictable change in the value of $\epsilon°$, whereas localization of the solute can lead to a change in its apparent A_s value [Eq. (14)]. However, Eq. (8) does not take into account the *interaction* of these two effects as in Fig. 3c and e. Therefore, for the case of polar (i.e., localizing) solutes and solvents, a term Δ_l must be added to Eq. (8):

$$\log(k_1/k_2) = \alpha' A_s (\epsilon_2 - \epsilon_1) + \Delta_l \tag{29}$$

The term Δ_l corrects Eq. (8) for the interaction of solute and solvent localization, and its resulting effect on k'. From our discussion of Fig. 3, it is clear that Δ_l should depend upon both the nature of the solute (X) and the mobile phase (i); Δ_l will be larger for increasing localization of both X and i. We therefore expect that Δ_l will be a function of parameters Δ_X (solute) and m_i (mobile phase); Δ_X measures the relative localization of X, and m_i increases with both the degree of localization of some mobile-phase solvent (e.g., C) and with its relative coverage of the adsorbent surface (θ_C). A linear-free-energy relationship between Δ_l and the parameters Δ_X and m_i is expected, because Δ_l' is a free-energy term which is the result of the *interaction* of effects produced by the localization of solute and solvent; such a linear-free-energy relationship has been experimentally verified (*13, 18*):

$$\Delta_l = -\Delta_X m_i \tag{30}$$

Because solvent–solute localization leads to decreased retention of the solute, the term Δ_l is negative.

The solvent selectivity parameter m of Eq. (30) is of primary interest in terms of controlling separation. For the case of a mobile phase A/C where the solvent C is localizing, the value of m is determined by the polarity of pure C ($m°$) and by the mole fraction θ_C in the monolayer:

$$m = m° f(\theta_C) \tag{30a}$$

The function $f(\theta)$ varies from 0 to 1 as θ goes from 0 to 1, and approaches a maximum value near $\theta_C = 0.75$. The solvent parameter m (for pure C) increases with the polarity of pure C or the value of ϵ_C.

Consider next two solutes X and Y, and mobile phases B and C [cf. Eqs. (26a) and (26b)]. From Eqs. (29) and (30) with $\epsilon_1 = \epsilon_2$ we can write

$$\log \alpha_c = \log \alpha_b + (\Delta_y - \Delta_x)(m_c - m_b) \tag{31}$$

If we select the mobile phase A/B to have $m_c = 0$, then we obtain the general relationship[15]

$$\log \alpha = C_1 + C_2 m \quad (31a)$$

The constants C_1 and C_2 are defined by the particular pair of solutes (X, Y) and by the adsorbent (e.g., silica).

The above model [Eqs. (30a) and (31a)] can be generalized to the case of mobile phases containing more than two solvents (*18*). It is also found that the parameter C_2 of Eq. (31a) is constant for a particular mobile-phase combination (e.g., A/C or A/B/C), but some variation of C_2 is possible between various localizing solvents C, D, etc. This latter effect, which is less important than the variation of α with m, is referred to as *solvent-specific solvent–solute localization* (*18*), and is discussed in Section III,B,3.

2. *Hydrogen-Bonding Systems*

The effect of solvent–solute hydrogen bonding on solvent selectivity can be considered in terms of the example of Section II,C. There we examined the case of a solute XH and a solvent C that can hydrogen-bond to form the complex XHC.

Now consider a second solute Y which is not a proton donor, and a second solvent B, which is not a proton acceptor. Because solute–solvent hydrogen bonding is not possible for the combinations XH/B, Y/B, or Y/C, corresponding equilibrium constants K^{bx}, K^{bx*}, K^{cy}, K^{cy*}, K^{by}, and K^{by*} all equal zero. We can define adsorption constants K_{xb}, K_{yc}, and K_{yb} analogous to K_{xc} of Eq. (23), whereupon as in the derivation of Eq. (24):

$$k_{xb} = (V_a/V_n)K_{xb} \quad (32a)$$

$$k_{yc} = (V_a/V_n)K_{yc} \quad (32b)$$

$$k_{yb} = (V_a/V_n)K_{yb} \quad (32c)$$

The separation factors $k_x/k_y = \alpha$ for the two mobile phases A/C and A/B can be defined as previously: α_c and α_b.

Consider first the situation where $K^{cx*} = K^{cx} = 0$; i.e., hydrogen bonding of XH and C can be ignored. Then Eqs. (28) and (31) yield

$$\log \alpha_c = \log \alpha_b + \alpha'(A_x - A_y)(\epsilon_2 - \epsilon_1) + (\Delta_y - \Delta_x)(m_c - m_b) \quad (33)$$

[15] Note that although m_C increases for pure solvent C as ϵ_C increases, for solvent mixtures the localization m of the mobile phase can be varied independently of $\epsilon°$, by varying the solvents B, C, etc. (and their proportions) in the mobile phase.

Finally, for nonzero (positive) values of K^{cx*} and K^{cx}, Eqs. (25) and (33) give

$$\log \alpha_c = \log \alpha_b + \underbrace{\alpha'(A_x - A_y)(\epsilon_b - \epsilon_c)}_{\text{(i)}} + \underbrace{(\Delta_y - \Delta_x)(m_c - m_b)}_{\text{(ii)}} + \underbrace{\log\left[\frac{1 + K^{cx*}\theta_C}{1 + K^{cx}N_C}\right]}_{\text{(iii)}} \quad (34)$$

Equation (34) is a general expression which accounts for the three known contributions to solvent selectivity in LSC: (i) solvent strength, (ii) solvent–solute localization (including solvent-specific localization), and (iii) solvent–solute hydrogen bonding. *The second term (ii) is generally the most important and most general contribution to solvent selectivity.*

The contribution of solvent–solute hydrogen bonding to selectivity [term (iii) in Eq. (34)] can be generalized for the presence of more than one basic or proton-acceptor solvent in the mobile-phase mixture:

$$(\log \alpha_c)_{iii} = \log \frac{1 + \sum^{i} K^{ix}\theta_i}{1 + \sum^{i} K^{ix}N_i} \quad (35)$$

For a given (reference) mobile phase A/B with $m_b = 0$, Eq. (34) can be simplified to

$$\log \alpha_c = c_a + c_2(\epsilon_c - \epsilon_b) + c_c m_c + \log\left[\frac{1 + K^{cx*}\theta_C}{1 + K^{cx}N_C}\right] \quad (36)$$

III. RESULTS AND DISCUSSION

A. Displacement Model—Non-Hydrogen-Bonding Systems

First consider LSC systems where hydrogen bonding between solute and solvent molecules cannot occur (no proton-donor molecules).

1. Validity of Eq. (8)

Equation (8) states that plots of log k' for a solute versus the parameter $\epsilon°$ (for all mobile phases) will yield a single straight-line curve. Equation (8) further predicts that the slope of this plot will be proportional to the cross-sectional area (A_S) of the solute molecule. These restrictions on the displacement model are considerably more stringent than those imposed by other correlational equations referenced in the Introduction (*2–10*). For example, the use of a single mobile-phase parameter ($\epsilon°$) allows the comparison in one plot of data from mobile phases which comprise mixtures of different solvents: A/B, A/C, A/B/C, etc. The use of a single mobile-phase parameter also allows the plotting of data for different solutes so as to span a wide range in mobile phase polarity—equivalent to the use of retention data for a single solute over a range in k' of three or more orders of magnitude (for experimental reasons, most studies vary k' by a factor of no more than 10–20).

The linearity of plots of log k' versus ϵ° has been demonstrated in many studies referenced in Ref. *1*. For example, Fig. 4 (from Ref. *1*) shows plots for 9 different solutes, using 7 different solvents (pentane, CCl_4, benzene, CH_2Cl_2) and/or mixtures thereof. Similar plots for an amino-phase column (*17*) are shown in Fig. 5 for 17 different solutes and hexane/tetrahydrofuran binary mobile phases.

Likewise, the Soczewinski equation (*2*) is based on the proportionality of ϵ° and log N_B for binary-solvent mobile phases A/B, when the concentration of B in the mobile phase is large enough so that $\theta_B \sim 1.0$. Since the Soczewinski equation has been confirmed for numerous LSC systems using silica as adsorbent (e.g., *22–27*), this constitutes confirmation of Eq. (8) as well.

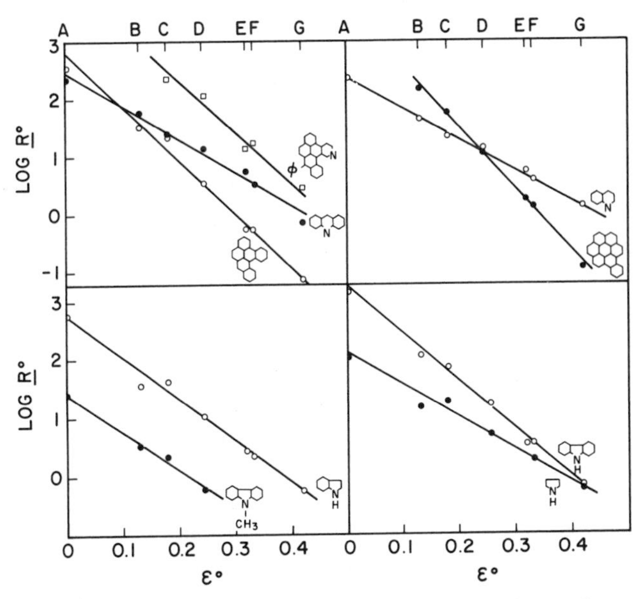

FIG. 4. Plot of log R (equivalent to log k') versus ϵ° for various solutes on alumina. Aromatic solutes are indicated. Mobile phases A–G (see top of figure) are as follows: A, pentane; B, 10:90 (v/v) CH_2Cl_2/pentane; C, CCl_4; D, 25:75 (v/v) CH_2Cl_2/pentane; E, benzene; F, 50:50 (v/v) CH_2Cl_2/pentane; G, CH_2Cl_2. Taken from Ref. (*1*).

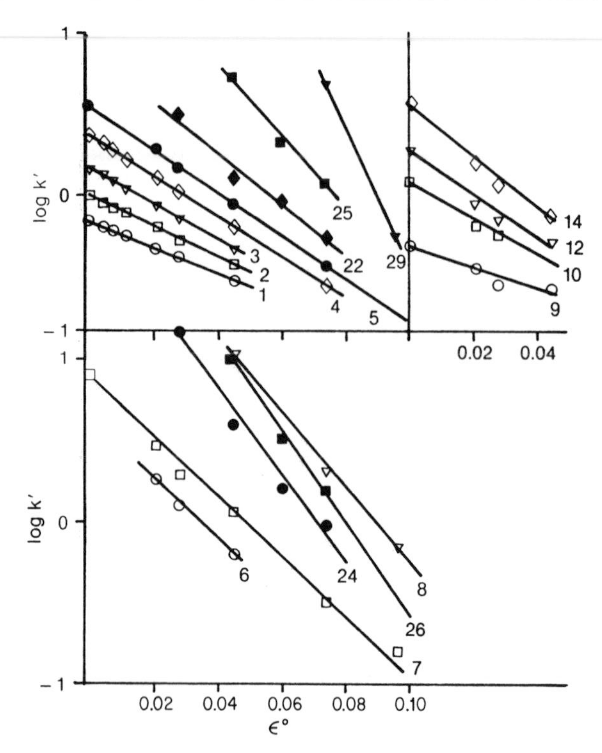

FIG. 5. Plot of log k' versus $\epsilon°$ for various solutes on an amino-bonded-phase column (*17*). Mobile phases are tetrahydrofuran/hexane mixtures; solutes are substituted aromatic compounds. Taken from Ref. (*17*).

The expected proportionality of log k' versus $\epsilon°$ plots with the solute A_s value has also been demonstrated in numerous LSC systems. For example, Fig. 6 shows a plot of experimental A_s values (slopes of log k' versus $\epsilon°$ plots) versus values calculated from the solute molecular dimensions, for various nonlocalizing solutes on an amino-bonded-phase column (aromatic hydrocarbons and ethoxylated nonylphenols). Another study (*34*) with alumina as adsorbent examined the ratio (k_1/k_2) [Eq. (8)] for various pairs (1 and 2) of pure solvents as mobile phase: pentane/CCl_4, CCl_4/benzene, benzene/CH_2Cl_2, and 52 different solutes (aliphatics, aromatics, polar, nonpolar). The deviation of individual experimental and calculated values of A_s was 1.5 units (1 standard deviation) for the range $5.1 \leq A_s \leq 21.2$.

For silica and amino-bonded phases as adsorbent, the experimental A_s value of Eq. (8) is larger than the calculated value (molecular size) for solute molecules that localize. This is summarized by Eq. (14), where

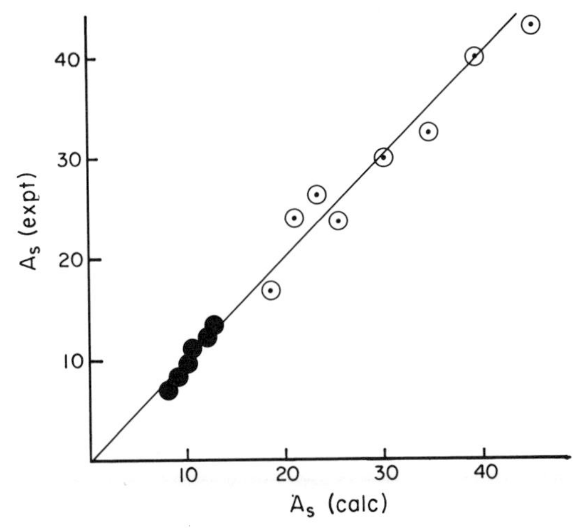

FIG. 6. Experimental versus calculated Values of A_s for nonlocalizing solutes on amino-bonded-phase (*17*): (●) aromatic hydrocarbons; (○) ethoxylated nonyl phenols; system of Fig. 5. Data replotted from Ref. (*17*).

$f_1(X)$ increases with increasing polarity and increasing localization of the solute X. For silica and amino-bonded phases, the quantity $f_1(X)$ can be related to the individual polar substituent groups **k** of the solute molecule

$$f_1(X) = \sum^{n} f_1(k) \tag{37}$$

That is, $f_1(k)$ for each solute group **k** increases with the adsorption energy of $\mathbf{k}(E_{ka})$, and the quantity $f_1(X)$ for the solute molecule is the sum of individual values of $f_1(k)$ for the various groups **k** in the molecule of X. Since we have ascribed values of $f_1(X)$ to the site-competition delocalization of the solute X, this implies that each group **k** in the solute molecule can localize initially. This in turn implies that intramolecular delocalization [Eq. (16)] of polar groups **k** in a solute molecule with more than one such substituent should not occur; i.e.,

$$R_M = \sum^{n} \Delta R_M \tag{38}$$

Hammers *et al.* (*35, 36*) have quantitated the relative contributions of these two effects in terms of the parameters β (intramolecular delocalization) and γ (site-competition delocalization). Combining their data for different adsorbents with corresponding data from (*1*), we have the following tabulation:

	Adsorbent		
Parameter	Alumina	Silica	Amino phase
β	1.0	0.5*, 0.4	0.3
γ	0.0	0.6*, 1.0	1–1.1

Here the asterisk indicates C_{18}-deactivated silica. There is a regular trend from left to right: as intramolecular delocalization (value of β) decreases, site-competition delocalization (value of γ) becomes more important. This is predicted by the above discussion. Furthermore, *there seems to be a regular transition in the nature of surface sites in the order alumina : C_{18}-silica : silica : amino-bonded phase*. This will be discussed further when we consider localization as a function of the nature of adsorbent sites (Section III,A,3).

2. Solvent Strength versus Mobile-Phase Composition

Values of ϵ° for the mobile phase are given by Eqs. (11)–(11b). We will consider three successively more complex cases, all involving binary-solvent mobile phases: (1) a mobile phase A/B, where A is nonpolar and B is polar but nonlocalizing, (2) a mobile phase A/C, where C is polar and localizing, and (3) a mobile phase B/C. Later, mobile phases containing more than two solvents will be discussed.

Case 1 is fairly straightforward, corresponding to the classic displacement model with no localization effects. For binary-solvent mobile phases A/B, Eq. (11a) can be solved for the *apparent* (experimental) value of ϵ_B in the mobile-phase mixture, given experimental values of ϵ° from Eq. (8) or Eq. (14). If localization and solute–solvent interaction effects can be ignored, then values of ϵ_B should be constant for various mixtures of A and B (N_B varying). It is instructive to plot experimental values of ϵ_B measured in this manner against the adsorbent surface coverage by B(θ_B). The quantity θ_B can also be derived from experimental values of ϵ° for the mixture A/B and various solvent parameters:

$$\theta_B = K_{ba}N_B/(N_A + K_{ba}N_B) \tag{39}$$

where K_{ba} is given by Eq. (10a).

Values of ϵ_B and θ_B measured as above from experimental ϵ° values for various nonlocalizing-solvent LSC systems are plotted in Fig. 7. The solvent B is variously CCl_4, benzene, or CH_2Cl_2; the adsorbent is silica or alumina. Similar results are obtained (*14*) for other nonlocalizing solvents: $CHCl_3$, 1- or 2-chloropropane, etc. As expected, values of ϵ_B are reasonably constant (±0.01–0.02 unit) for most cases. The somewhat greater scatter of the top plot (CH_2Cl_2/alumina) might be attributed to mobile-phase interactions for larger values of θ_B (and N_B), except that a similar

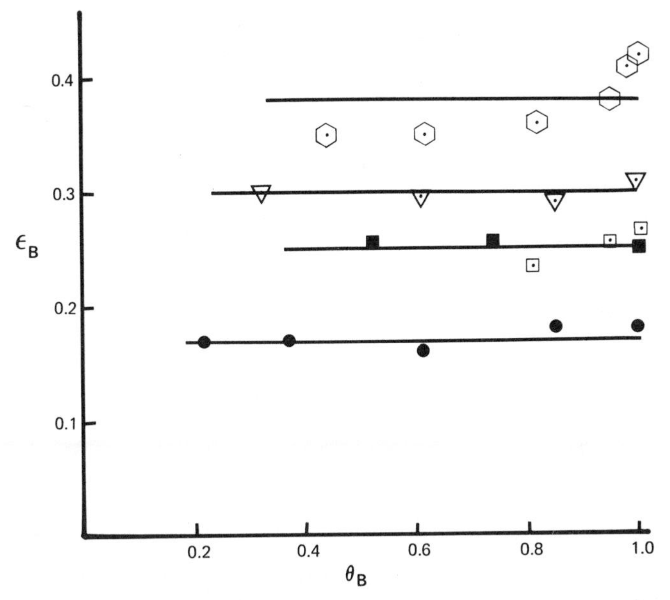

FIG. 7. Dependence of B-solvent strength ϵ_B on coverage of adsorbent surface θ_B. Mixtures A/B of nonpolar solvent A and polar nonlocalizing solvent B: ⊙, CH_2Cl_2/pentane/alumina; ▽, CH_2Cl_2/pentane/silica; ⊡, benzene/hexane/silica; ■, benzene/pentane/silica; ●, CCl_4/pentane/alumina. Data taken from Snyder and Glajch (*14*) and Hara *et al.* (*25*).

effect would then be predicted in the next lower plot of Fig. 7 (CH_2Cl_2/silica)—but is not found. In any case, a uniform pattern is observed for these ϵ_B versus θ_B plots for nonlocalizing solvents B: ϵ_B is constant for all values of θ_B.

Case 2, involving mobile phases A/C where C is localizing, is examined in Fig. 8 in similar fashion. Here a regular departure from the constant ϵ_B values of Fig. 7 is seen. For relatively nonpolar solvents such as isopropyl ether (bottom plot), there appears to be a slight increase in ϵ_C as θ_C is decreased. This effect is progressively more pronounced as the polarity of the C-solvent (and ϵ_C) is increased, as seen in Fig. 8 for the following increasingly polar C-solvents: ethyl ether, acetone, and isopropanol. This effect has been discussed in detail (*14*),and arises from restricted-access delocalization of the C-solvent with increasing coverage of the adsorbent surface (as in Fig. 1b).

Equation (12) describes the variation in ϵ_C and θ_C as a function of parameters ϵ'_C and ϵ''_C, and a function of θ_C: $\%_{lc}$ (percent localization of C). This relationship is reasonable in terms of the description of restricted-access delocalization given earlier. That is, a significant change in

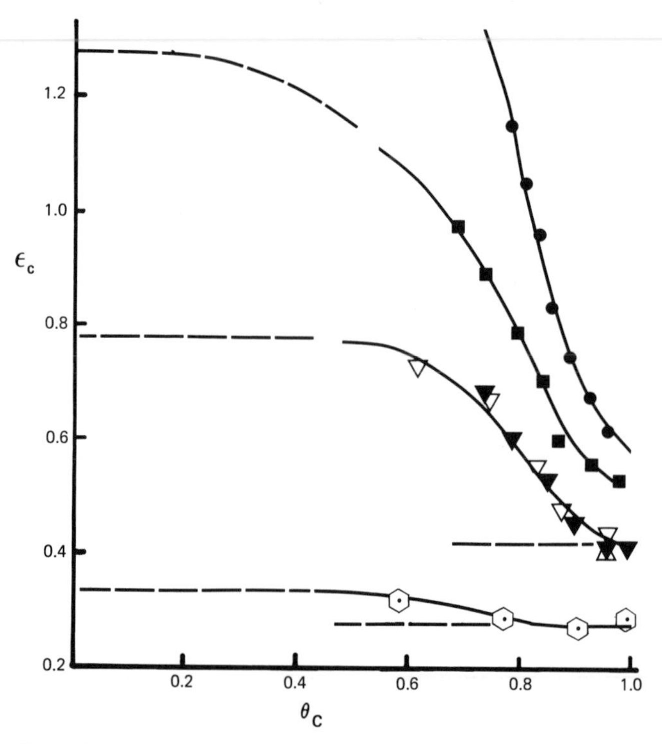

FIG. 8. Dependence of C-solvent strength ϵ_C on coverage of adsorbent surface θ_C. Mixtures A/C of nonpolar solvent A and polar localizing solvent C: ●, isopropanol/hexane/silica; ■, acetone/hexane/silica; ▼, ethyl ether/hexane/silica/; ▽, ethyl ether/pentane/silica; ⊙, isopropyl ether/pentane/alumina. Data taken from Snyder and Glujch (*14*) and Hara *et al.* (*25*). Curves through data are best fit to Eqs. (12) and (40).

C-solvent localization should occur at a surface coverage ($\theta_C \sim 0.75$) which is the same for all C-solvents. Data for ϵ_C versus θ_C reported in Ref. *14* suggest that the function $\%_{lc}$ can be approximated well by the empirical fitting expression

$$\%_{lc} = (1 - \theta_C)\,[1/(1 - 0.94\theta_C) - 14.5\theta_C^9] \tag{40}$$

Several mobile phases A/C and alumina as adsorbent showed close agreement with Eqs. (12) and (4), as summarized in Fig. 9a.

The similar behavior of ϵ_C versus θ_C plots for silica is shown in Fig. 9a,b, where data for several C-solvents (ethyl ether, ethyl acetate, acetone, isopropanol) in admixture with hexane or pentane are plotted as experimental values of $\%_{1c}$ [best fit to Eq. (12)] versus experimental values of θ_C. The data for isopropanol (solid circles) are displaced to the right

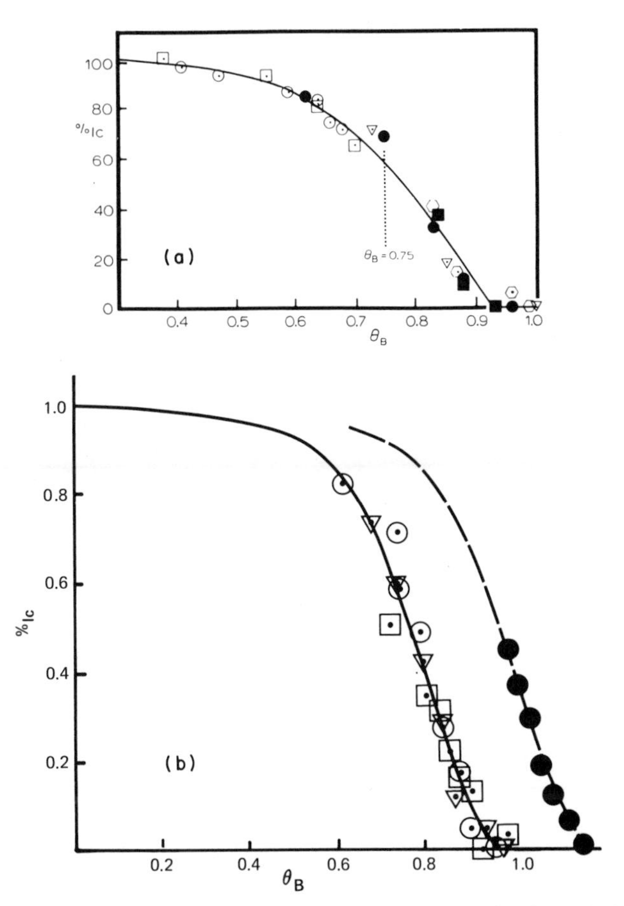

FIG. 9. Similarity of C-solvent delocalization on silica and alumina. (a) Alumina: (⊙) acetonitrile, (⊡) acetone, (▽) ethyl acetate. Silica: (●) diethyl ether, (■) ethyl acetate. (b) Silica: (⊙) ethyl ether/hexane, (⊡) ethyl acetate/hexane, (▽) acetone/hexane, (●) isopropanal/hexane.

by 0.2 unit in θ_C for better observation. It is clear that the calculated curve of Fig. 9b [from Eq. (40)] provides a good fit to these data, showing that solvent delocalization on silica is essentially similar to that observed for alumina (Fig. 9a). The use of Eqs. (11a), (12), and (40) then allows the calculation of ϵ° for any binary A/C, given the solvent parameters summarized in Table I. This overall procedure is summarized in the Appendix. Agreement between experimental and calculated values of ϵ° for mobile phases A/B and A/C has been found equal to about ± 0.015 unit in ϵ° (*14*, *16*), for a few hundred mobile phases and $0 \leq \epsilon^\circ \leq 0.66$.

TABLE I

Solvent Properties of Interest in Liquid–Solid Chromatography

		Solvent strength						
		Silica		Alumina		Selectivity $m°$		
Solvent	n^b	ϵ'	ϵ''	ϵ'	ϵ''	Silica	Alumina	$f_l(C)/n_b$ (silica)
Freon FC-113[a]	—	0.02	0.02			—	—	—
Hexane, heptane[a]	—	0.00	0.00	0.00	0.00	—	—	—
Carbon tetrachloride	5.0	0.11	0.11	0.17	0.17	−0.35[c]	−0.09	0.0
Toluene	6.8	0.22[b]	0.22[b]	0.30	0.30	−0.43[c]	−0.16	0.0
1-Chloropropane[a]	3.8	0.28	0.28	0.31	0.31			0.0
2-Chloropropane[a]	3.8	0.28	0.28	0.31	0.31	−0.23[c]	0.02	0.0
Benzene	6.0	0.25	0.25	0.32	0.32	−0.42[c]	−0.15	0.0
Chlorobenzene	6.7	—	—	0.31	0.31	−0.12[c]	0.12	0.0
Bromoethane	3.8	—	—	0.34	0.34	−0.16[c]	0.08	0.0
Chloroform[a]	5.0	0.26	0.26	0.36	0.36	0.10	0.34	0.0
Methylene chloride[a]	4.1	0.30	0.30	0.40	0.40	0.10	0.29	0.0
Isopropyl ether[a]	5.1	0.42[b]	0.32[b]	0.31	0.28			0.8[d]
Triethyl amine[a]	6.2	—	—	0.85	0.36	0.65[c]	0.82	
Ethyl ether[a]	4.5	0.78	0.43	0.50	0.38	0.43	0.62	1.0
Ethyl sulfide	5.0	—	—	0.43	0.39	−0.03[c]	0.29	
1,2-Dichloroethane[a]	4.8	0.36[b]	0.34[b]	0.47	0.44	0.14[c]	0.35	0.0
MTBE[a]	4.1	1.01	0.48	—	—	0.82	0.77[c]	1.1
Tetrahydrofuran[a]	5.0	1.00	0.53	0.76	0.51	0.65[c]	0.82	1.3
Acetone	4.2	1.00	0.53	1.01	0.58	0.87	1.02	1.3
Ethyl acetate[a]	5.2	0.94	0.48	0.77	0.60	0.60[c]	0.77	1.6
Dioxane[a]	6.0	1.00	0.51	0.79	0.61	—	—	1.0
Pyridine	5.8	—	—	0.95	0.70	1.09[c]	1.22	—
Deithyl amine[a]	4.0	—	—	1.8	0.71	—	—	—
Acetonitrile[a]	3.1	1.00	0.52	1.31	0.55	1.19	1.31	1.6
Isopropanol[a]	4.4	1.8	0.60	—[e]	0.82	—	—	4.1
Methanol[a]	3.7	2.1	0.7	—[e]	0.95	—	—	—

[a] A preferred solvent for LSC with UV detection.
[b] Estimated value; see Eq. (12) and the Appendix.
[c] Value estimated via Eq. (41).
[d] Estimated value; see Eq. (15).
[e] Eq. 12 does not apply.

The similar analysis of values of ϵ_B and ϵ_C (as in Figs. 7–9) for amino-phase packings is interesting for several reasons. Since the latter adsorbent is much weaker than alumina or silica, the adsorption terms E^*_{xa} and E^*_{ma} of Eq. (4) will be smaller, and the mobile-phase terms E_{xn} and E_{mn} will be relatively more important. As a result, Hennion *et al.* (*37*) have suggested that interactions in solution need to be considered in any precise description of retention in these LSC systems.[16] The amino-phase packing is also radically different from alumina and silica in terms of adsorption site characteristics. As we will see, this leads to an absence of restricted-access delocalization (which occurs for *rigid* adsorbents such as alumina and silica). Figure 10 shows experimental plots of ϵ_B (or ϵ_C) versus θ_B for several mobile phases A/B.[17] These each show a rather similar behavior, rising at low values of θ_B as in Fig. 8, but *also* rising as θ_B approaches a value of 1.0. The apparent increase in ϵ_B as θ_B approaches unity has been attributed to mobile-phase interaction effects (*17*), as discussed in Section II,C. The rise in ϵ_B at low values of θ_B is superficially similar to that caused by restricted-access delocalization (Fig. 8), but differs in major respects, as discussed next.

There are two reasons to suspect that restricted-access delocalization is not responsible for the rise in ϵ_B at small θ_B in Fig. 10. First, an *apparent* $\%_{lc}$ function can be derived from these latter data and is plotted in Fig. 11. A single curve (solid line) fits the data points for different B-solvents, as expected for restricted-access delocalization. However, this solid curve in Fig. 11 differs dramatically from the dashed curve of Fig. 11, which is the plot of $\%_{lc}$ versus θ_B for localizing solvents (C) on alumina and silica [Eq. (40)]. Whereas the latter curve suggests localized adsorption of the B-solvent for θ_B as large as 0.75 ($\%_{lc} = 0.5$), the corresponding value of θ_B for the localized monolayer on an amino-phase surface is only 0.18. It is reasonable to assume that localized solvent molecules might occupy as much as 75% of the adsorbent surface, but unreasonable to conclude that a localized monolayer *cannot* exceed 18% of a monolayer. Furthermore, the same fractional coverage of the surface by localized solvent molecules (at saturation or large θ_B) is found for both silica and alumina, in agreement with the common physical explanation that applies to restricted-access delocalization on both adsorbents: physical crowding of adjacent adsorbed molecules, so that completion of a localized monolayer is impossible. On this basis, restricted-access delocalization on an amino-phase surface should also be associated with 75% coverage of the amino-phase surface by localized solvent molecules at saturation. Since this is not

[16] One of the reviewers has pointed out that mobile-phase terms will have the same absolute significance, regardless of the magnitude of adsorbed-phase interactions. This is correct, in terms of their relative effects on separation. However, the *observable* effects of mobile-phase interactions on experimental k' values will be more apparent when these are no longer masked by large adsorbed-phase interactions (which are only approximately describable by simple theories such as the one given in this chapter).

[17] Note the expanded scale of Fig. 10 as compared to that of Figs. 7 and 8, which allows us to magnify the dependence of ϵ_B on θ. However, at the same time this similarly magnifies experimental imprecision (the lines through each point of Fig. 10 indicate the precision of these experimental ϵ_B values, ± 1 standard deviation).

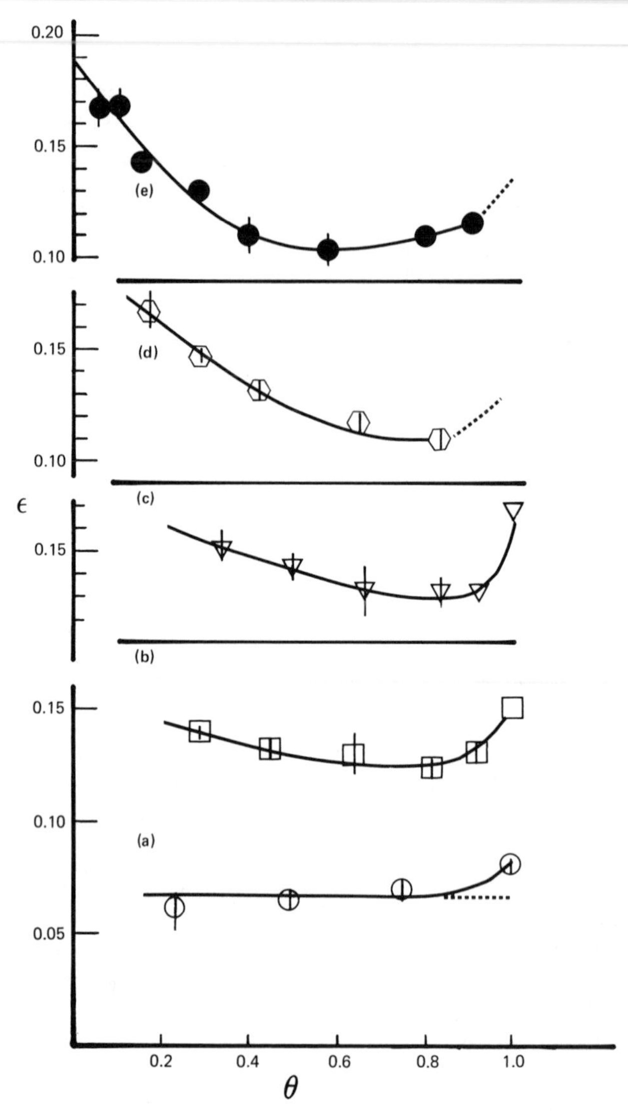

FIG. 10. Dependence of solvent strength of B- or C-solvent (ϵ) on its coverage of the surface (θ) for amino-silica. Mixtures A/B or A/C, where A is hexane [data taken from Snyder and Schunk (*17*)]: ○, CCl_4; □, CH_2Cl_2; ▽, $CHCl_3$; ⬡, ethyl acetate; ●, tetrahydrofuran.

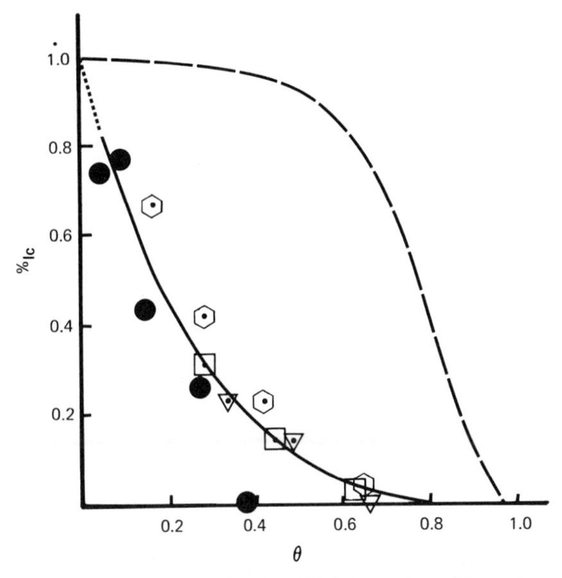

FIG. 11. Apparent B-solvent delocalization ($\%_{lc}$) for amino-silica column as a function of B-solvent surface coverage (θ): (---) Eq. (40); data of Fig. 10.

observed (Fig. 11), we can conclude that restricted-access delocalization does not occur for adsorption onto amino-phase adsorbents.

A second reason to doubt the occurrence of restricted-access delocalization on amino-phase columns is that this increase in ϵ_B (Fig. 11) is observed for both localizing and nonlocalizing solvents: ethyl acetate and tetrahydrofuran (localizing) and CH_2Cl_2 and $CHCl_3$ (nonlocalizing). This contrasts with theory, which predicts that delocalization effects are only associated with localizing compounds.

An alternative explanation for Fig. 11 is more attractive. Thus, Hammers *et al.* (*35*) have shown that partially silanized silicas behave as a deactivated silica, with reduced surface polarity, but with the same ability as silica to adsorb polar solutes. If the coverage of this particular amino-phase packing surface is incomplete (which is generally the case for packings prepared by the procedure used for amino-phase materials), then two different classes of adsorption sites exist: residual silanols and the alkylamino groups (—R—NH_2). The silanols are inherently more polar and therefore stronger adsorption sites (cf. *36*). Therefore polar solvents B or C will be preferentially adsorbed onto these silanols, rather than onto the more numerous alkylamino groups. If this is the case, Fig. 11 suggests that about 18% of the total surface sites are accessible silanols, whereas the balance are alkylamino groups.

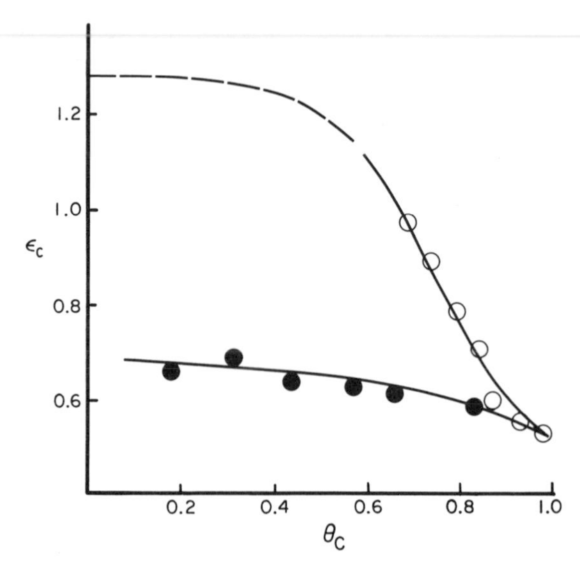

FIG. 12. Dependence of C-solvent strength ϵ_C on coverage of adsorbent surface θ_C. Mixtures B/C (●) and A/C (○), where C is acetone, B is benzene, and A is hexane (silica adsorbent). Data taken from Snyder and Glajch (*16*).

The data of Fig. 11 thus indicate that restricted-access delocalization does not exist for amino-phase packings. Furthermore, the importance of the variation of ϵ_B with θ_B, shown in Fig. 11, is insignificant, so far as calculations of ϵ° are concerned. We will therefore assume that ϵ_B is constant for a given solvent B and amino-phase packings, in the calculation of ϵ° for multicomponent mobile phases (Appendix).

Case 3 involves the use of mobile-phase mixtures B/C, where B and C are polar, but only C can localize. The solvent B can also be a blend of a nonpolar solvent A with a pure solvent B (A/B), in which case we deal with ternary-solvent mobile phases A/B/C.[18] The effect of restricted-access delocalization on values of ϵ_C in mixtures B/C is illustrated in Fig. 12 (solid circles) for mobile phases composed of acetone (C) and benzene (B), with silica as adsorbent. The open circles are comparative data for acetone/hexane (A/C) mixtures. The same tendency of ϵ_C to increase for smaller θ_C is found for acetone in mixtures B/C as is found for mixtures A/C. This is further confirmed by the data of Fig. 13, where $\%_{lc}$ for several mobile phases B/C (B is benzene, C varying) are plotted versus θ_C. The precision of these experimental values of $\%_{lc}$ is much reduced in the case of mobile phases B/C in comparison to those of type A/C, as can be seen in

[18] More complex (e.g., four or more solvents) mobile phases are treated in Ref. *16*.

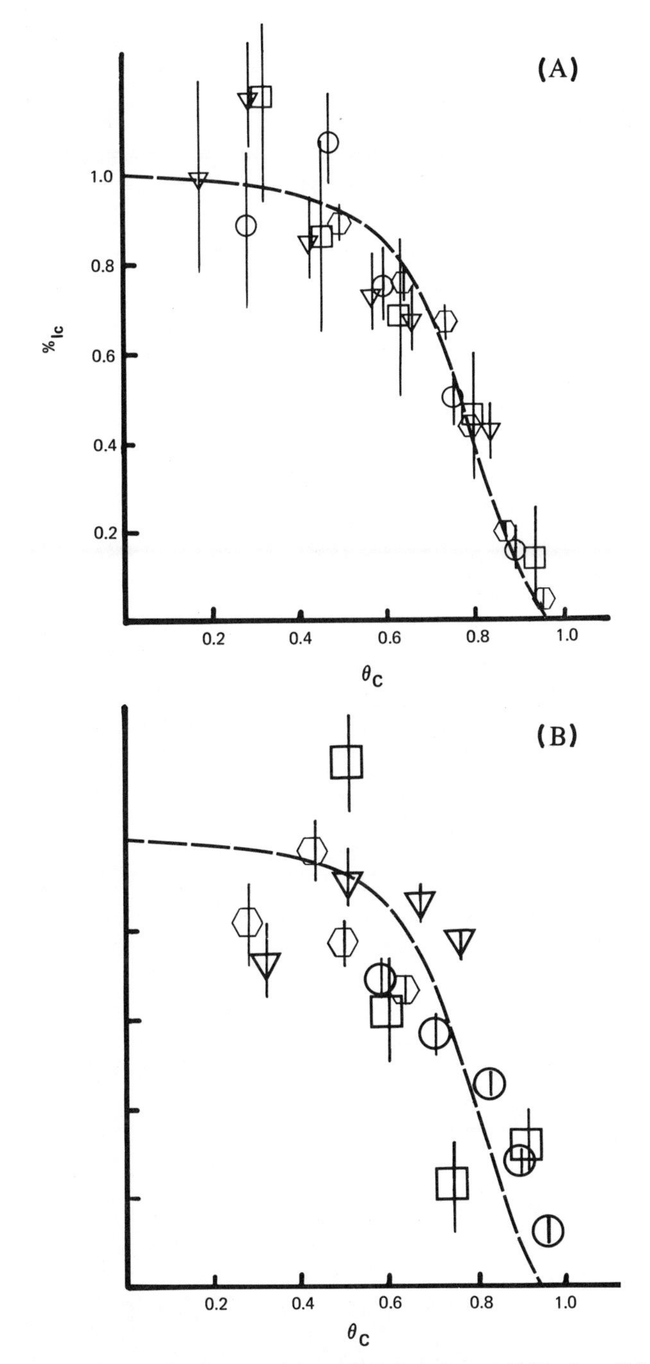

FIG. 13. C-solvent delocalization (as $\%_{lc}$) on silica for mixtures B/C, where B is benzene ⊙ (A) ○, ethyl ether; □, ethyl acetate; ▽, acetone; ⊙, tetrahydrofuran; (B) □, acetonitrile; ▽, isopropanol; ⊙, ethanol. Data taken from Snyder and Glajch (*16*).

the plots of Fig. 13. One reason is that the ratio ϵ'/ϵ'' is reduced for mixtures B/C, as discussed below. The second reason is that a given error or imprecision in an experimental value of ϵ° creates up to a 3-fold greater error in values of ϵ_C for mixtures B/C in comparison to corresponding mixtures A/C. Each of these effects translates into a larger error in $\%_{lc}$ as a result of some error in ϵ° for mixtures B/C as compared to A/C.

The vertical lines through each point of Fig. 13 correspond to the experimental uncertainty in $\%_{lc}$ for an error of ± 0.005 unit in the associated ϵ° value (a reasonable estimate of the precision of experimental ϵ° values). In the case of the less polar C-solvents of Fig. 13a, all data points appear to fit the dashed curve [Eq. (40)] with acceptable precision. The fit for the more polar C-solvents of Fig. 13b is poorer, and may reflect mobile-phase interactions as discussed in Section II,C. However, the overall trend in ϵ_B' versus θ_B is clearly that predicted by restricted-access delocalization.

Returning to Fig. 12, it is seen that the value of ϵ_C' for a mixture B/C is smaller than in a mixture A/C, as predicted by Eq. (15). This is the result of site-competition delocalization (superimposed onto restricted-access delocalization), the same phenomenon that leads to increase in the A_s value of localizing solute molecules, as compared to the value calculated from the molecular dimensions of the solute. The function $f_l(C)$ of Eq. (15) is the same function as $f_l(X)$ in Eq. (14) for delocalization of solute molecules. A previous study (Fig. 3 of Ref. *16*) has shown that plots of $f_l(C)$ and $f_l(X)$ versus the adsorption energy Q_k° of the solute or solvent substituent k that is localized (E_{ka}) give a single curve through points for both solvents (C) and solutes (X). This function $f_l(C)$ is tabulated in Table II and can be used to estimate values of ϵ_C' for mobile phases B/C, when the experimental value of $f_l(C)/A_c$ is not known for the solvent C (see Table I).

Site-competition delocalization with resulting decrease in ϵ_C' for stronger solvents B in mixtures B/C [Eq. (15)] has been observed for silica as adsorbent (*16*) and several solvents C; this effect should also occur for

TABLE II

Values of f_1(C) for Determining ϵ' When Mobile Phase A/C Has $\epsilon_A \neq 0.00$ [Eq. (15)]

Q_k°	f_l(C)	Q_k°	f_l(C)
0.0	0.0	3.0	4.5
0.5	0.0	4.0	6.5
1.0	0.4	5.0	8.5
1.5	1.5	6.0	10.5
2.0	2.5	8.0	14.5

amino-phase packings, in view of the observed localization of polar solutes on this adsorbent. Although only limited data are available for site-competition delocalization of the solvent in LSC on alumina, one study (*15*) allows the calculation of ϵ_C values for dioxane as C-solvent and alumina as adsorbent, for ternary-solvent mobile phases A/B/C. Unfortunately, resulting values of ϵ'_C are not very precise due to the use of this more complex mobile phase. However, some improvement in precision can be obtained by averaging ϵ'_C values for mobile phases where the $\epsilon°$ value of the mixture A/B is similar. The result of such averaging is shown in the following tabulation:

$\epsilon°$ for A/B			
Range	Average	Average ϵ'_C	Mobile phases[a]
0.22–0.23	0.224	0.90 ± 0.16	#1–3
0.28–0.32	0.310	0.89 ± 0.21	#4–7
0.38	0.381	0.89 ± 0.08	#8–10

[a] From Table I of Ref. *15*.

While the precision of the resulting ϵ'_C values is poor, it appears (as predicted) that the value of ϵ'_C is independent of the strength of the mobile phase exclusive of C (A/B). That is, site–competition delocalization of the C-solvent does not occur on alumina.

3. *Localization Effects as a Function of Adsorption Sites*

Three types of solvent or solute delocalization have now been examined, as summarized in Table III for three different adsorbent types (four, if we distinguish C_{18}-deactivated silica from silica). The theoretical requirements on the configuration and density of adsorption sites were discussed earlier (Section II,B) for a given type of localization/delocalization to be possible. In each case the nature of adsorption sites is fairly well understood for the four adsorbents of Table III, as disucssed in Ref. *1* and *17* and shown in Fig. 14. Thus, in the case of alumina, surface hydroxyls do not function as adsorption sites. Although surface oxide atoms are capable of interacting with acidic adsorbate molecules (see below), in most cases the adsorbate will interact with a cationic center (either aluminum atom or lattice defect) in the next layer. As a result, we can say that in most cases adsorption sites on alumina are buried within the surface, rather than being exposed for covalent site–adsorbate interaction. These sites are also rigidly positioned within the surface. Finally, the

TABLE III

Localization–Delocalization Effects for Different Adsorbents. The Nature of Adsorption Sites in Each Case

	Alumina	C_{18}-Silica	Silica	Amino phase
Delocalization effect				
Intromolecular[a]				
Solute (β)	1.0	0.5	0.4	0.3
Site-competition[a]				
Solute (γ)	0.0	0.6	1.0	1.0–1.1
Solvent	Absent	?	Present	
Restricted-access	Yes	?	Yes	No
Nature of sites				
Positions fixed on Surface?	Yes	Yes	Yes	No
Sites exposed (above surface)?	No	Partial	Yes	Yes
Concentration of sites (μmol/m^2)	High	Low (4–6)	High (8)	Low (2)

[a] Numbers are values of β or γ defined in text.

concentration of alumina sites is believed to be comparable to that on the silica surface, although they cannot be measured directly.

In the case of silica, the active sites are surface silanols. This means that the sites are exposed for direct interaction with adsorbate molecules. The concentration of these sites (~8 μmol/m^2) is large compared with the likely concentration of adsorbate molecules in the surface. C_{18}-silica is

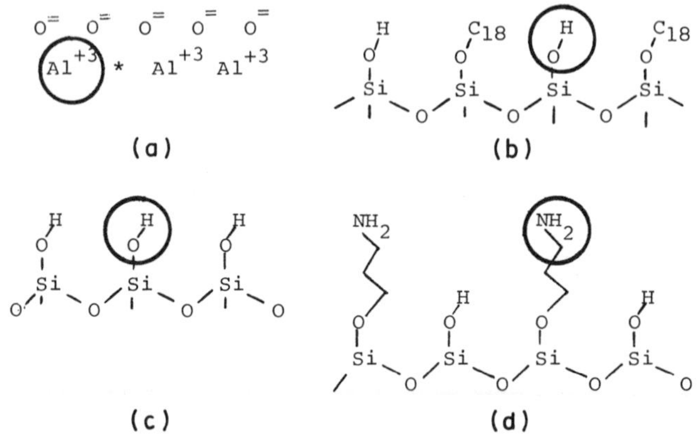

FIG. 14. Representation of the surface structure of four LSC adsorbents; the active site is circled in each case: (a) alumina; * refers to vacancy-site, (b) C_{18}-silica, (c) silica, (d) amino phase.

similar to silica, the only difference being a lower concentration of accessible silanol sites. In both cases, these silanols are rigidly positioned on the surface.

Terminal amino groups are believed to comprise adsorption sites on the surface of amino-phase packings (*17*). These (1) are exposed for direct interaction with adsorbate molecules, (2) are not rigidly positioned on the surface, and (3) have a much lower surface concentration (2–3 μmol/m^2). These characteristics of the various LSC packings so far discussed are summarized in the surface representations shown in Fig. 14 for each adsorbent. The major adsorption sites in each case are indicated by enclosure within a circle (except the vacancy site for alumina, shown as an asterisk in Fig. 14).

Let us next examine the various delocalization phenomena for these four adsorbents in terms of the properties of their surface sites and the earlier theoretical discussion. Consider first *intramolecular* delocalization of the solute. From theory we expect this effect to be most pronounced for adsorbents with rigidly positioned sites, for adsorbents with "buried" rather than exposed sites, and for adsorbents with a lower concentration of surface sites. The parameter β measures the relative importance of intramolecular delocalization. The data of Table III show that alumina exhibits this phenomenon in the greatest degree. This is reasonable in terms of the rigid positions and unexposed nature of these sites (despite their higher concentration). Amino-phase packings exhibit intramolecular delocalization to the least degree, which arises from the presence of nonrigidly positioned and exposed sites on the surface. Silica shows in-between behavior, because the sites are rigidly positioned as compared with amino-phase sites. The slightly greater β value for C_{18}-silica versus silica may be due to the lower concentration of surface sites and/or the less exposed nature of these sites in the case of C_{18}-silica. However, the difference in β values is not experimentally significant.

Site-competition delocalization will be favored by surface sites that are accessible to lateral interaction by solvent molecules that are adjacent to a localized solute or solvent molecule. The relative accessibility of sites for this type of interaction is as follows: alumina (least), C_{18}-silica, and silica or amino phase (most). This is in fact the order of increasing site-competition delocalization noted in Table III for delocalization of either solvent or solute molecules.

One exception to the absence of site-competition delocalization on alumina is noted for solutes with acidic substituents. These are held quite strongly on alumina, and it is believed this is due to the direct interaction of these proton-donor groups with surface oxide groups on alumina. Thus, for this situation the requirements for site competition are met, and the effect on $(A_s)_{expt}$ is in fact observed. For example, various pyrroles exhibit A_s values on alumina which are larger than calculated from molecular dimensions by about 2.6 units (*1*), and this effect vanishes when the N—H group is replaced by N—CH_3 (e.g., lower left-hand portion of Fig. 4, where *N*-methyl carbazole has about the same $(A_s)_{expt}$ value or slope as for indole).

Restricted-access delocalization is favored by a high concentration of surface sites and by the rigid positioning of these sites on the surface. Alumina and silica meet these requirements and exhibit restricted-access delocalization. Amino-phase columns do not meet these requirements and do not exhibit this phenomenon. C_{18}-silica would be expected to exhibit the effect in lesser degree, due to the lower concentration of surface sites.

The varying behaviors of these various adsorbents in terms of these delocalization effects can therefore be explained in terms of structural differences in active sites. Presumably the same logic can be extended to other types of adsorbents, as well as used to advantage to design new adsorbents of predetermined characteristics. Delocalization effects should become less pronounced as the surface energy of the adsorbent decreases, and values of E_{xa} and E_{ma} become smaller.

4. *Interactions among Solvent and Solute Molecules. Effect on* $\epsilon°$

For non-hydrogen-bonding solvents and solutes, it has been argued earlier that the effect of intermolecular interactions on $\epsilon°$ may be small. This hypothesis can now be examined in terms of deviations of LSC retention data (and values of $\epsilon°$) from the displacement model described to this point. First, consider less polar mobile phases of the type A (nonpolar, nonlocalizing) or B (polar, nonlocalizing). Because intermolecular interactions of molecules *i* with the surrounding liquid phase *j* are proportional to the polarities of *i* and *j*, values of γ_{ij} for mobile-phase systems of this type (A/B etc.) will be relatively small. In one study (*34*), for 52 solutes and the pure-solvent mobile phases pentane, CCl_4, benzene, and CH_2Cl_2, experimental slopes of log k' versus $\epsilon°$ plots exhibited an average deviation from calculated slopes, equivalent to ± 1.5 units in A_s (avg. $A_s = 10$). Since the average difference in $\epsilon°$ for the two solvents 1 and 2 [Eq. (8)] in that study was 0.14 unit, this implies an uncertainty in $\epsilon°$ of ± 0.02 unit (1 standard deviation). Note that these studies involve pure-solvent mobile phases, with none of the complexity associated with mixed-solvent mobile phases (different solvent composition in two phases). Furthermore, this discrepancy between experimental and calculated solvent strengths for individual compounds also reflects other approximations in the displacement model, apart from intermolecular interactions.

The effect of intermolecular interactions between solvent and solute

TABLE IV

Values of ϵ° and ϵ^m for Pure-Solvent Mobile Phase[a]

Solvent	ϵ°	ϵ^m	$\epsilon^m - \epsilon^\circ$
CCl_4	0.067	0.081	0.014
CH_2Cl_2	0.124	0.150	0.026
$CHCl_3$	0.130	0.170	0.040
Ethyl acetate	0.110	—	—
Tetrahydrofuran	0.115	0.125	0.01

[a] Amino-silica column (*17*).

molecules should be more easily distinguishable, when the (primary) molecule–surface interactions E_{ia} are weaker. This is the case for the less retentive amino-silica packing (*17*), where ϵ_B and ϵ_C values appear to increase with increasing concentrations of B or C in the mobile phase (Fig. 10 and related discussion). The data of Fig. 10 then suggest the values of ϵ° and ϵ^m (Eq. 21) for the pure-solvent mobile phase shown in Table IV. Because of differences in the (arbitrary) values of α' for the above amino-phase adsorbent (1.00) and for silica (0.57), the above molecular-interaction contributions to ϵ° (i.e., $\epsilon^m - \epsilon^\circ$) should be reduced by about half for comparison with silica ϵ° data. This suggests contributions to ϵ° from intermolecular-interaction contributions of as much as 0.02 unit for solvents as polar as $CHCl_3$. If these intermolecular interactions are ignored, the experimental measurement of apparent ϵ° values (ϵ^m) via Eq. (8) with averaging of the resulting ϵ_B values for a solvent B will include an average of the log Q term of Eq. (19). In practical terms, this means that the contribution of these intermolecular interactions to variation in ϵ° should be roughly one-half of the maximum possible value (i.e., in mobile phases A/B with N_B small, $Q \approx 1$), or up to 0.01 unit in ϵ° for solvents as polar as $CHCl_3$ and ethyl acetate with amino-silica as packing.

The preceding examples suggest an uncertainty in ϵ° of the order of ± 0.01 unit (for amino-silica) as a result of intermolecular interactions. Looking again at Fig. 10, we see that this contribution (from Q) occurs at θ_B values close to unity, or N_B values above 0.5. The assessment of the importance of this effect in polar-solvent LSC systems with silica or alumina is more difficult, because according to Fig. 8 there is also a contribution to values of ϵ° from restricted-access delocalization of the solvent. For very polar B-solvents with large A_m values, however, N_B can be varied substantially (with variation in Q) while keeping θ_B close to unity and avoiding change in ϵ_B due to variation in solvent delocalization. Some data from Ref. *14* for dioxane/pentane and pyridine/pentane (alumina as adsorbent) are of interest in this regard (see Table V):

TABLE V

Data for Dioxane/Pentane and Pyridine/Pentane (Alumina)

Mobile phase	N_B	θ_B	ϵ_B	ϵ° Expt.	ϵ° Calc.	Standard deviation in ϵ°
Dioxane/pentane	0.130	0.97	0.60	0.372	0.383	
	0.310	0.99	0.60	0.472	0.479	
	0.573	1.00	0.61	0.549	0.547	
	1.000	1.00	0.63	0.630	0.610	±0.015
Pyridine/pentane	0.056	0.96	0.71	0.378	0.363	
	0.137	0.98	0.67	0.435	0.465	
	0.322	0.99	0.70	0.565	0.566	
	0.588	1.00	0.73	0.662	0.637	±0.024

[a] Taken from Ref. *14*.

These polar-solvent systems where θ_B (and solvent localization) remain essentially constant show relatively constant values of ϵ_B and good agreement between experimental and calculated ϵ° values (±0.020 overall). The implication is that the Q term of Eq. (21) is introducing at most 0.01–0.02 unit of uncertainty into values of ϵ°. This in turn implies a maximum value for Q of the order of 1.5[19] for these various LSC systems. These estimates of Q are also in accord with the ability of Eq. (11a) to reliably predict values of ϵ° for binary-solvent mobile phases (±0.015 for a wide range of solvents A and B—Refs. *14* and *16*). Whereas Q seems to vary by no more than 30% in non-hydrogen-bonding systems, values of γ_{xn} and γ_{mn} can vary by factors of up to 10 or more (*32*). *There is obviously considerable cancellation of these γ_{ij} terms in Eq. (18a), as previously discussed.*

B. Solvent Selectivity Effects—Non-Hydrogen-Bonding Systems

1. Solvent-Strength Selectivity

Solvent selectivity effects in LSC are accounted for by terms (i), (ii), and (iii) of Eq. (34). We will first discuss solvent-strength selectivity term (i) and solvent–solute localization selectivity term (ii), leaving hydrogen-bonding selectivity term (iii) to the following section. As already indicated, solvent-strength selectivity term (i) is of limited value in optimizing retention in LSC. This effect is directly based on the validity of Eq. (8),

[19] That is, assume $A_s = 15$, $\alpha' = 0.6$, then Eq. (21) gives $\log Q = 0.02 \times 15 \times 0.6 = 0.18$.

which has been widely documented. No further discussion of term (i) is therefore required.

2. Solvent–Solute-Localization Selectivity

The effect of solvent–solute localization and its effect on selectivity is given by Eq. (31a):

$$\log \alpha = C_1 + C_2 m \tag{31a}$$

This relationship has been experimentally verified for numerous mobile phases (and different solvents) and a wide variety of solutes, with both alumina (*13*) and silica (*18*) as adsorbents. Some examples of the applicability of Eq. (31a) in these LSC systems are given in Fig. 15a (alumina) and Fig. 15b (silica). The use of 3-solvent or 4-solvent mobile phases (*18, 20*) allows the continuous variation of m while holding $\epsilon°$ constant, which greatly facilitates retention optimization by maximizing α without changing $\epsilon°$.

The dependence of m on θ_C for a localizing solvent C in mobile phases A/C or A/B/C is given by Eq. (30a). According to our theory of how these solvent–solute-localization effects arise (i.e., competition of localizing solvent and solute molecules for the same site), we expect that $f(\theta_C)$ will increase slowly with θ_C until the maximum localized-coverage is approached ($\theta_C > 0.5$). The function $f(\theta_C)$ should then increase sharply and level out for $\theta_C > 0.75$. Figure 16 shows actual data for $f(\theta_C)$ in the case of alumina (Fig. 16a) and silica (Fig. 16b). The same function $f(\theta_C)$ is shown in each plot, and it is apparent that experimental data fall close to the solid curves (best-fit values of $m°$ for each solvent C). The shape of the solid curve is as predicted, based on the analysis of Fig. 3.

The parameter Δ_x should increase in magnitude with increasing localization of the localizing group k in the molecule X. Since localization increases with the adsorption energy Q_k° of the group k, it is expected that values of Δ_x should increase with the value of Q_k° for the compound X. This correlation was tested (*13*) with alumina as adsorbent, and is shown in Fig. 17 [Δ_x in Eq. (30) is equivalent to $-\Delta°$ here]. As expected, there is a regular increase in Δ_x with increasing Q_k°, providing further validation of the present theory of the origin of these solvent-selectivity effects. However, for two compounds with similar k' values, the values of Q_k° are often not very different; the scatter in the plot of Fig. 17 therefore suggests that predictions of the quantity $(\Delta_x - \Delta_y)$ in Eq. (31) will not be very precise. This is generally the case in practice, so that *ab initio* predictions of values of C_2 (based on the molecular structures of X and Y) for Eq. (31a) are often not practical.

In favorable cases involving greatly different values of Q_k°, solvent–

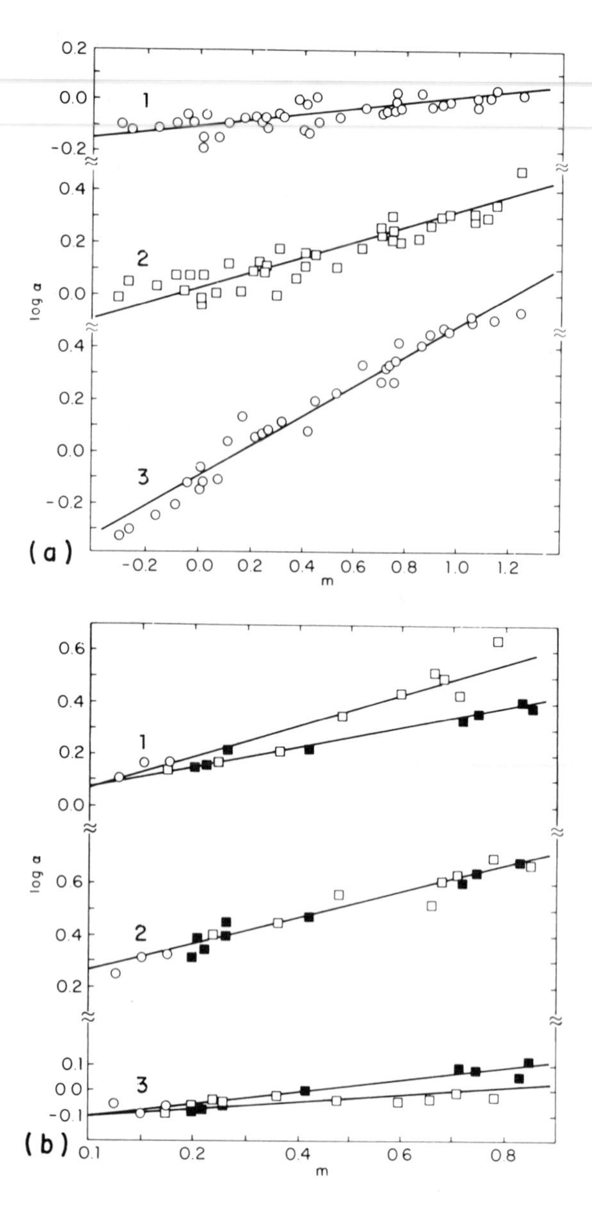

FIG. 15. Correlation of solvent selectivity effects with solvent–solute localization and Eq. (31a): (a) alumina, different solute pairs for each plot, 18 different polar solvents: (b) silica, as in (a), except ternary- and quaternary-solvent mixtures used; localizing agent: □, MTBE; ■, ACN; (○) chloroform or dichloromethane. The solute pairs are as follows: (a) 1—1-naphthaldehyde and 1-cyanonaphthalene; 2—1-nitronaphthalene and 1,2-dimethoxynaphthalene; 3—1,5-dinitronaphthalene and 1-acetylnaphthalene; (b) 1—1-nitronaphthalene and 2-methoxynaphthalene; 2—1,5-dinitronaphthalene and 1,2-dimethoxynaphthalene; 3—methyl 1-naphthoate and 2-naphthaldehyde. Reprinted from Snyder *et al.* (*13, 18*).

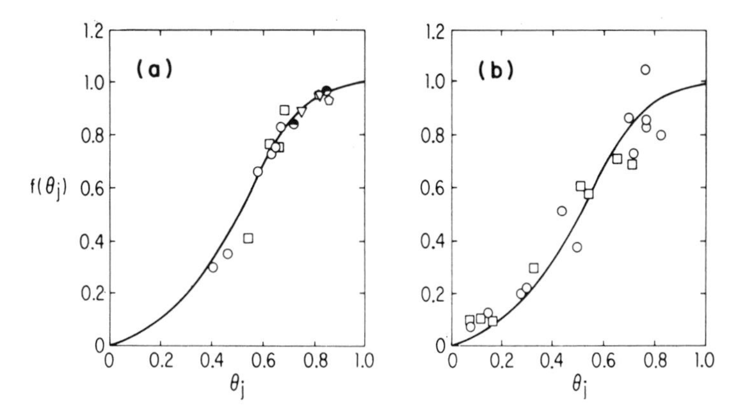

FIG. 16. Variation of solvent selectivity parameter m with surface coverage θ of localizing solvent [Eq. (30a)]: (a) variation of mobile-phase m values of $f(\theta)$ for alumina ○, pyridine; □, acetone; ▽, THF; ◓, ethyl acetate; (b) same for silica ○, MTBE; □, ACN. Reprinted from Snyder *et al.* (*18*).

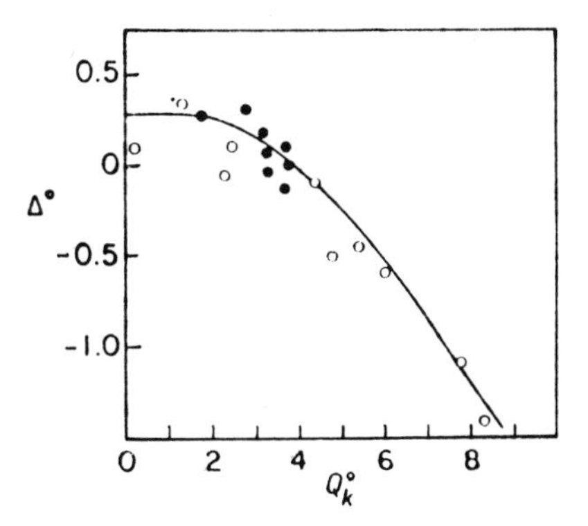

FIG. 17. Dependence of solute-localization parameter Δ_x [Eq. (30)] on adsorption energy Q_k° of localizing group k in solute molecule. Reprinted from Snyder *et al.* (*13*) for alumina as adsorbent, with $\Delta^\circ = -\Delta_x'$.

solute selectivity can provide considerable control of α values through manipulation of mobile-phase composition. Figure 15 shows changes in α of as much as fivefold for the separation of these substituted-naphthalene solutes. In extreme cases, changes in α of several hundred-fold are possible. One example (*13*) is worth citing, for trinitrobenzene (TNB) and *N,N*-dimethylnaphthoic acid amide (DMNA):

Solvent	m	k'[a] TNB	k'[a] DMNA	α
Benzene	−0.4	0.32	88	275
5% (v/v) acetonitrile/ 10% (v/v) benzene/pentane	0.8	5.9	7.1	1.2
Q_k°		2.8	8.3	
Δ° (= $-\Delta_x$)		0.56	−1.40	

[a] Adsorbent: 4.3% H_2O Al_2O_3.

While such large changes in α are limited to compounds of quite different structure, it is likely that some preparative separations can be greatly facilitated by maximizing α in this way without limit. In analytical separations, however, values of α larger than 5–10 can actually be detrimental, because they result in too large a difference in retention times for the two solutes (the "general elution problem," Ref. *1*).

So far we have stressed the ability of solvent–solute localization to facilitate the separation of functionally different compounds: those with localizing groups k that are not the same. However, similar differences in Q_k° can be found for the same functional group substituted into different isomers. For example, Palamareva *et al.* (*38*) reported the separation of numerous diastereomeric solute pairs on silica and found striking changes in α for different mobile-phase compositions. These data have since been correlated (*39*) in terms of solvent–solute localization, and it was found that Eq. (31a) provides an excellent fit to these data. This is shown in Fig. 18 for several of these diastereomer pairs, where log α is plotted against calculated values of m for the four mobile phases studied in Ref. *38*. Not only are the data accurately predicted by Eq. (31a), but changes in α by a factor of up to 2.5 are possible as a result of changing mobile-phase composition and m, even for these isomeric compounds of closely related structure.

Since the origin of solvent–solute localization on various adsorbents arises from a single general phenomenon, we should expect that m° values for various solutes will be correlated among adsorbents of similar group selectivity (similar Q_i° values for different solute groups i; see Fig. 10.10 of Ref. *1*). Therefore, a plot of m° values for silica should correlate with similar values for alumina. This is observed, as shown in Fig. 19. The straight line shown in Fig. 19 is given by

$$m^\circ \text{ (silica)} = -0.25 + 1.1m^\circ \text{ (alumina)} \tag{41}$$

Equation (41) allows the estimation of m° values for silica, based on the more extensive set of known values for alumina, as summarized in Table I.

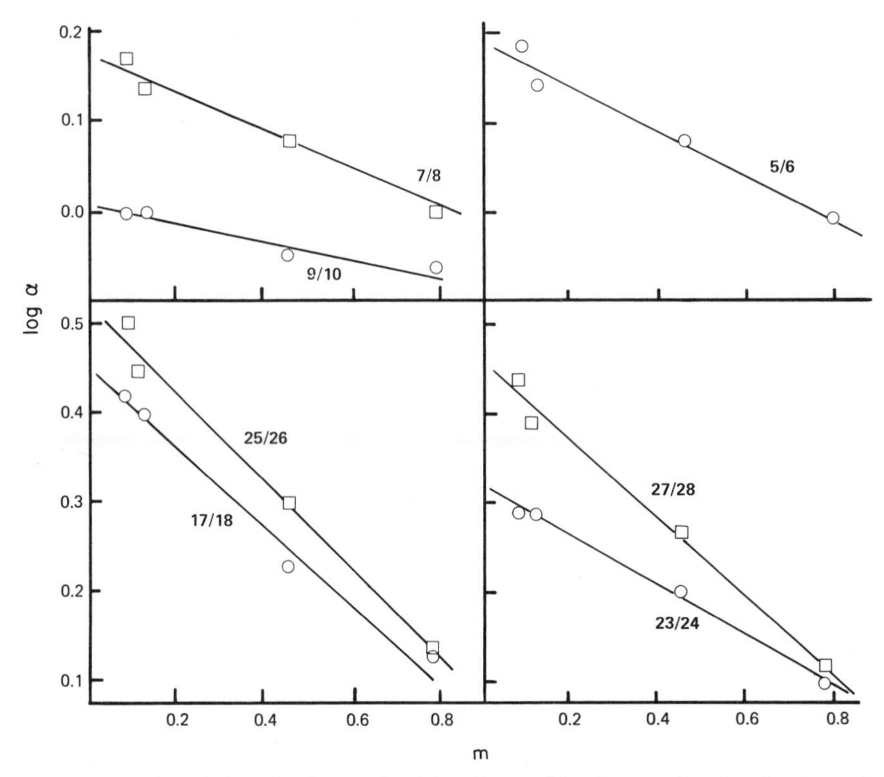

FIG. 18. Correlation of solvent selectivity effects with solvent–solute localization and Eq. (31a). Data for selected diastereomeric solute pairs. Reprinted with permission from Snyder (*39*).

3. Solvent-Specific Localization Selectivity

The data plots of Fig. 15b (silica) are differentiated for the use of methyl-*t*-butyl ether (MTBE) or acetonitrile (ACN) as localizing solvent C in the mobile phase. It is seen that for some solute pairs (Fig. 15a and c) the open squares (MTBE) fall on a different curve than the closed squares (ACN). This implies that the constant C_2 in Eq. (31a) is solvent-specific, rather than being constant for all solvents (as first-order theory would predict). A similar behavior is observed for alumina as well. Figure 15a plots data for 18 different polar solvents B or C, and some scatter of these plots of log α versus m is observed here, as in Fig. 15b for silica. The variation of C_2 with the localizing solvent C used for the mobile phase has been shown (*18*) to correlate with the relative basicity of the solvent, or its placement in the solvent classification scheme of Refs. *40* and 41. Thus, for relatively less basic solvents (groups VI or VII in Refs. *40* and *41*),

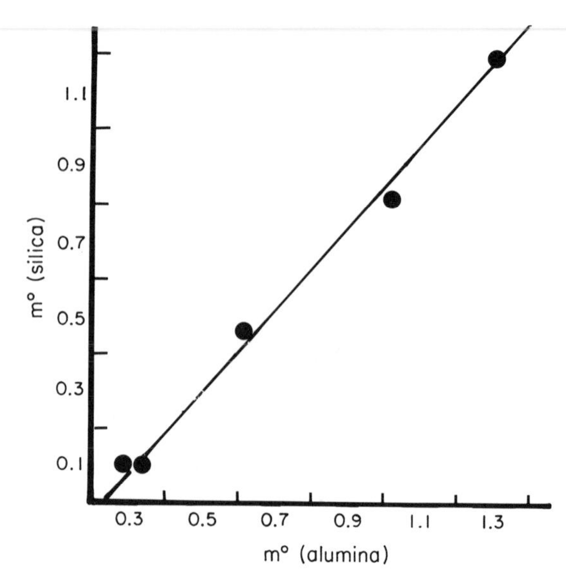

FIG. 19. Correlation of solvent-selectivity parameter m° for silica versus alumina. Values taken from Table I.

values of C_2 are quite constant. For more basic solvents (groups I or III in Ref. *40*), values of C_2 are more variable and generally different from values for less basic solvents. In Fig. 15b, ACN (group VI) and MTBE (group I) are maximally different, and deviation of the data from a single curve then results. Presumably, this *solvent-specific localization* (variable values of C_2) arises from the molecular "fine structure" present in the monolayer of localized C. It seems less likely that solute–solvent interactions are responsible, since each curve (data for ACN or MTBE) is linear in the localization parameter m.

Solvent-specific localization is generally only one-half to one-third as important as solvent–solute localization in affecting α values (*18*). However, its effects are nevertheless important in overall solvent-optimization strategies for LSC (see Section III,E).

C. Solvent Selectivity Effects—Hydrogen-Bonding Systems

Less is known about the relative importance of solvent–solute hydrogen bonding in affecting solvent selectivity, although several workers have postulated such effects (e.g., *1*, *42*, *43*). However, one must be cautious in accepting these various observations at face value, because solvent–solute localization effects will normally be large in systems that can exhibit solvent–solute hydrogen bonding. Thus, one must first quantitatively account for terms (i) and (ii) of Eq. (34), before a quantitative assessment of term (iii) is possible.

One such attempt at measuring term (iii) and correlating it with Eq. (34) has been reported (*19*) for the solute pair 1-naphthol/ 1-acetonaphthalene, with silica as adsorbent, and MTBE and ACN as localizing basic solvents. Naphthol is a proton donor (X–H), whereas acetonaphthalene (Y) is not. The constant C_2 of Eq. (36) (equal $A_x - A_y$) could be calculated, and experimental values of α_c were available. This allowed Eq. (36) to be expressed as

$$(\log \alpha_c)_{corr} = \log \alpha_c - C_2(\epsilon_c - \epsilon_b)$$
$$= C_a + C_c m_c + \log\left[\frac{1 + K^{cx*}\theta_C}{1 + K^{cx}N_C}\right] \quad (42)$$

In this case, the log α_c values were corrected to a reference solvent strength ($\alpha_b = 0.237$), then the latter values of $(\log \alpha_c)_{corr}$ were correlated (least-squares fit) to the right-hand side of Eq. (42). The resulting correlation gave $C_a \sim C_c \sim 0$, and values of K^{cx*} and K^{cx} that differed for the two solvents (MTBE and ACN): $K^{cx*} = 7$, $K^{cx} = 50$ (MTBE); $K^{cx*} = 11$, $K^{cx} = 25$ (ACN). The fit of experimental values of $(\log \alpha_c)_{corr}$ compared to the values calculated by Eq. (42) and the above-derived fitting constants is shown in Fig. 20. The resulting correlations are reasonable and show

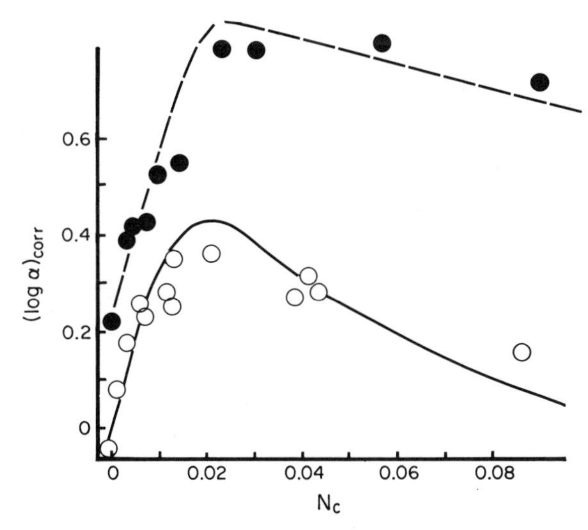

FIG. 20. Contribution of solvent–solute hydrogen bonding to separation selectivity (see text). The data are for MTBE (○) and ACN (●) as basic solvent. The ACN curve is displaced vertically by +0.2 log units. Reprinted from Snyder (*19*).

initial excess retention of the naphthol as N_C increases from a value of zero, followed by eventual decline in α as hydrogen bonding in the mobile phase becomes more important.

A major challenge to completing a practical model and description of mobile-phase effects in LSC is the further elucidation of hydrogen-bonding effects. This will involve a more fundamental classification of solutes and solvents in terms of their proton-donor and proton-acceptor properties, so that values of K^{cx} can be estimated as a function of the molecular structures of solute X and solvent C. It will also require a more precise description of the adsorbate-surface bonding that occurs in the adsorbed monolayer, so that values of K^{cx*} can likewise be rationalized and predicted.

D. Calculated versus Experimental Solvent Isotherms

Equation (39) allows the calculation of the surface mole fraction θ_B of B-solvent as a function of N_B, for mobile phases A/B (or A/C).[20] The weight $W°$ of B-solvent in the adsorbed monolayer at saturation ($\theta_B = 1$) can be calculated from the B-solvent A_b value and its molecular weight (see Refs. *1* and *14*). The uptake of B-solvent by the adsorbent for some value of N_B is then given as $\theta_B W°$. Several studies (*4, 12*) have reported experimental solvent isotherms (plots of $\theta_B W°$ versus N_B) for silica, and it is of interest to compare these data with isotherms calculated as above.

Figure 21 summarizes experimental isotherm data for several LSC systems (*4, 12*). In each case, the solid curve is calculated from Eq. (39). It should be noted that values of W_B° are calculated from the B-solvent molecular weight and dimensions, whereas the θ_B values are based on LSC retention data and derived values of ϵ_B, ϵ_A, n_b, etc. Thus, these calculated curves are not adjusted to the experimental data of Fig. 21—they represent totally independent calculations. This should be contrasted with other attempts at isotherm fitting, where up to three adjustable parameters are invoked to achieve as close a fit as possible.

Figure 21a–c are plots for nonlocalizing solvents: chloroform, benzene, and toluene. In the plots of Fig. 21a and b, there is a tendency of later points to fall below the calculated curve. Since the data must eventually converge to the curve at $\theta_B = 1.0$, this suggests simple experimental error in the values of $\theta_B W°$ for these later points. When the adsorbate is rather weakly adsorbed (as in this case), the determination of adsorbent uptake values ($\theta_B W°$) involves the subtraction of two large numbers, and the

[20] That is, from $K = 10^{\alpha' n_b(\epsilon_B - \epsilon_A)}$ (*14, 15*) and experimental values of ϵ_B as in Figs. 7 and 8.

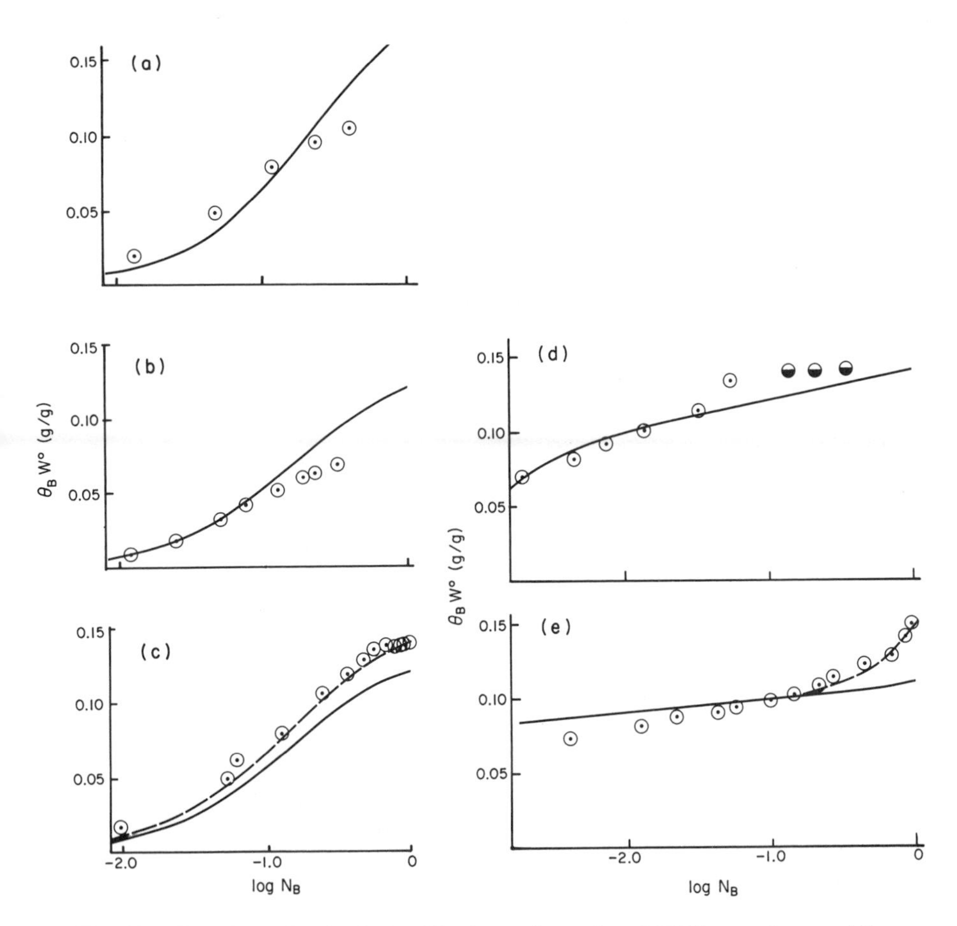

FIG. 21. Experimental and calculated isotherms for various LSC binary-solvent mobile phases /AB and silica as adsorbent. Solid curves are calculated from present model using parameters of Table I (based on retention data only); points are experimental uptake values from Scott and Kucera (*4*) and Snyder and Poppe (*12*). (a) $CHCl_3$/hexane; (b) benzene/hexane; (c) toluene/hexane; (d) ethyl acetate/hexane; (e) isopropanol/hexane.

results are correspondingly less accurate. In the case of the plot of Fig. 21c, the data points lie consistently above the calculated curve. While the adsorbent used in the plots of Fig. 21a–c is nominally the same (Partisil 10 silica), the data of Fig. 21a and b were determined in one laboratory (*4*), whereas the data of Fig. 21c were determined in a different laboratory (*12*). Thus, it is possible that small differences exist in the surface areas of the two adsorbents. The dashed curve of Fig. 21c is adjusted for a 15% increase in silica surface area, and is seen to give a good fit to the data.

Figure 21d and e show similar isotherm plots for two B-solvents that localize (ethyl acetate and isopropanol). The ethyl acetate data of Fig. 21d (circles) show reasonable adherence to the calculated curve, with the exception of some scatter at higher concentrations of ethyl acetate. Again, this may reflect experimental error.

The isopropanol isotherm in Fig. 21e shows reasonable adherence of data points to the calculated curve, except for values of $N_B > 0.2$, where the data points are seen to rise significantly above the calculated isotherm. This may reflect the buildup of a second adsorbed layer in this case, which is plausible in view of the ability of isopropanol molecules to self-hydrogen-bond. However, the extent of this second-layer buildup is never greater than about 35% of a monolayer.

On balance, the plots of Fig. 21 suggest that calculated values of θ_B from the present approach are reasonably close to actual isotherm values. Thus, these isotherm data can be regarded as supporting the present displacement model (and related equations), or at the least, not disproving the model. Whether the present approach can be extended to predict isotherm data with an acceptable accuracy for other purposes (e.g., preparative separations with column overload) remains to be seen. This will require careful studies of the same adsorbent sample, measuring both solvent isotherm data and appropriate solute retention values, with use of the solute retention data to derive solvent parameters for calculations of θ_B.

A final point which emphasizes the remarkableness of the correlations of Fig. 21 is that retention data used for these calculations were collected on different silicas in two laboratories, over the period 1962–1980 (*14, 24–27*). The actual isotherm data were obtained in two other laboratories on still a different silica sample.

E. Mobile-Phase Optimization Strategies

The overall approach to method development in LSC is as follows (e.g., *44*):

(1) Select a column; e.g., 15 cm in length, 5-μm particles of silica.
(2) Carry out a separation with an arbitrary mobile phase; e.g., pure solvent B in Fig. 22.
(3) Adjust solvent strength for the optimum range of k' values; e.g., 50:50 (v/v) B/A as in Fig. 22.

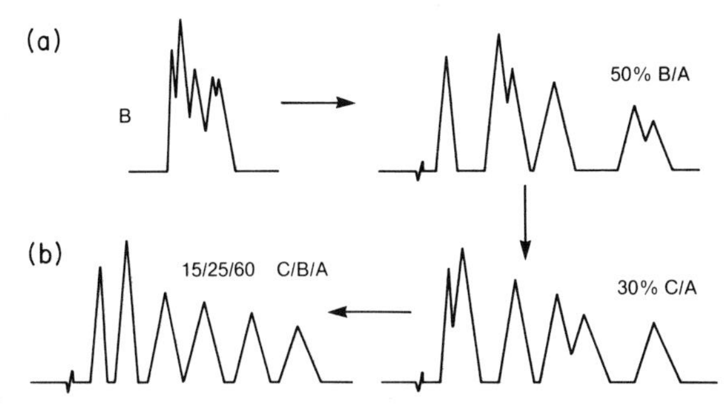

FIG. 22. Hypothetical example of retention optimization in liquid–solid chromatography. (a) Solvent strength k' is optimized; (b) selectivity α.

(4) If one or more solute pairs overlap severely, try another polar solvent C in place of B; select composition to have same strength as for 50 : 50 (v/v) B/A.
(5) If solute α values change by too much (leading to mixing of other bands) when solvent C is substituted for B (as in Fig. 22), "fine-tune" solvent selectivity by using ternary-solvent mixtures C/B/A (Fig. 22).

This is more or less the classical approach to selectivity optimization, except that in the past binary-solvent mobile phases were used most commonly.

On the basis of theory presented in this article it is possible to improve on the above approach to method development and retention optimization. But consider first the limitations of the classical approach. One problem is that of predicting the composition of the mobile phase A/C, such that its solvent strength is equal to that of 50:50 (v/v) B/A. The present discussion has shown that calculations of ϵ° are more complex than had previously been believed. Furthermore, calculations of solvent strength will be required for mobile phases containing more than two solvents (e.g., Fig. 22), and this has only recently become possible (*14–16*). As discussed in the Appendix, it is now possible to carry out such calculations of solvent strength versus composition, and to design mobile phases of equal solvent strength for use as in Fig. 22 (e.g., 30 : 70 (v/v) C/A and 15 : 25 : 60 (v/v/v) C/B/A). This procedure has been taken one step further, in that nomographs are now available for conveniently estimating the strength of useful multisolvent mobile phases in LSC on silica (*21*).

The second problem with classical selectivity optimization is in the choice of optimum solvents C, D, etc. for maximum change in selectivity or α values. Note that once large changes in α are effected, "fine-tuning" with ternary- or quaternary-solvent mixtures can be used for optimum spacing of bands within the chromatogram. The present discussion suggests that two major selectivity effects can be utilized in this selection process: solvent–solute localization (Section III,B,2) and solvent-specific localization (Section III,B,3). Thus, our initial objective should be to define solvents B, C, and D showing, respectively, minimum localization selectivity (B), maximum solvent–solute localization (C), and maximum solvent-specific localization (D). This can in turn be achieved by selecting (a) a solvent B with small value of $m°$, (b) a basic polar solvent C with large value of $m°$, and (c) a nonbasic polar solvent D with large value of $m°$. With these three solvents, plus a fourth nonpolar solvent A (e.g., hexane) to adjust solvent strength independently of solvent selectivity, we can achieve almost all of the selectivity that is possible with a given LSC adsorbent for a particular sample.

An optimum set of solvents A, B, C, and D has been chosen (*20*), based on hexane (A), methylene chloride (B), MTBE (C), ACN (D). Using the latter three polar solvents, we can define a selectivity triangle as in Fig. 23a. This triangle demapses the relative concentrations of mobile phase in terms of B, C, and D. Thus, the point at the top of the triangle corresponds to pure B, the point at the lower left of the triangle to pure ACN, and the point at the lower right of the triangle to pure MTBE.[21] To the extent that localization-selectivity effects dominate, all possible solvent selectivities can be created by some composition within the triangle of Fig. 23a. The compositions within the triangle of Fig. 23a change for different solvent strengths, having more or less hexane added to solvents B/C/D.

For a given solvent strength value, seven compositions within the triangle of Fig. 23a can be defined, as illustrated in Fig. 23b. Ideally, solvent selectivity (or log α) will change by a similar amount in going from any one of the compositions in Fig. 23b to an adjacent point in the triangle. This is discussed in Ref. *20*. Now the sample can be separated with each of the seven mobile phases defined as in Fig. 23b (see Ref. *21* for examples). Finally, the optimum mobile-phase composition can be selected by interpolating the data from these seven separations. This can be done visually in many cases, but computer programs can also be used to simplify and optimize this process—as we will shortly see.

For multivariable optimization as in the present case, so-called overlapping-resolution mapping (ORM) can be a useful approach to computer-assisted optimization of solvent selectivity (*20, 46*). This will be

[21] These three solvents are not completely miscible with hexane, but this problem can be avoided by substituting Freon-113 for hexane (*45*); hexane/ACN solutions of intermediate compositions are not miscible.

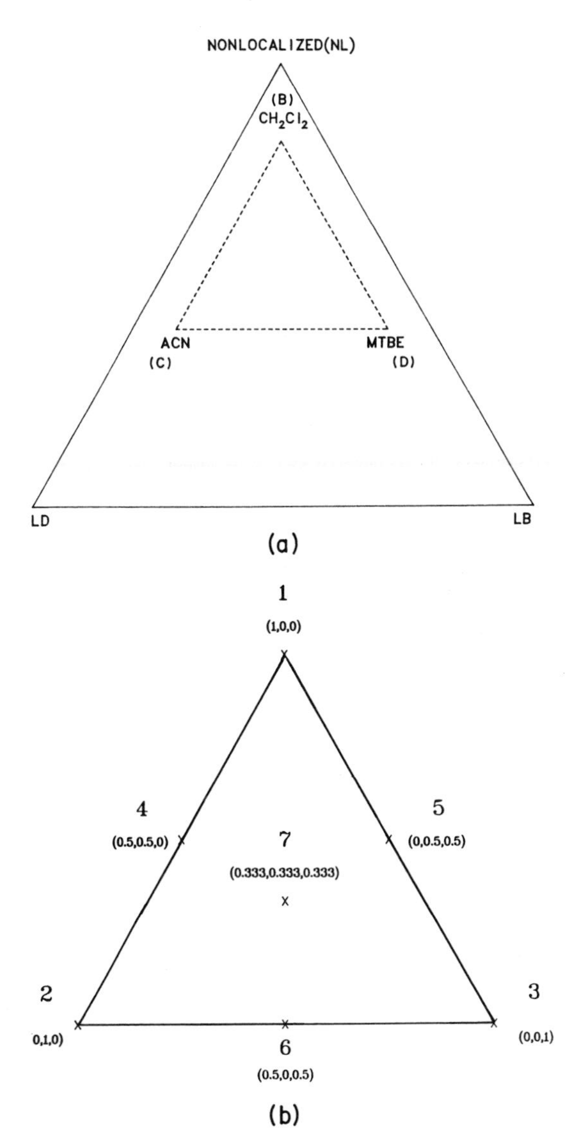

FIG. 23. The solvent selectivity triangle for separations on silica: LD, localized dipole; LB, localized base. (a) general form; (b) seven compositions for retention optimization. Reprinted with permission from Glajch *et al.* (*20*).

illustrated here for the separation of a 13-component mixture of substituted naphthalenes. Application of the scheme of Fig. 23 yielded the retention data of Fig. 24, for the seven compositions of Fig. 23b.

Application of a computer program to these data of Fig. 24 yielded calculated values of α for every possible solute pair in the seven mobile phases used. The program then interpolated these data over the entire compositional triangle by fitting to a quadratic curve, to yield values of α as a function of mobile-phase composition. Finally, these α values are plotted in trilinear form (Fig. 25) in such a manner as to indicate mobile-phase compositions of optimum selectivity. Figure 25a shows such a plot for band pair 6–8, where the white region indicates resolution of the two bands (on one 25-cm silica column) greater than the minimum desired ($R_s > 1.0$).

When this procedure is repeated for every possible pair of overlapping bands in the sample, and the R_s plots for each band pair are superimposed, the ORM plot of Fig. 25b results. Now it is seen that the white area for $R_s > 1.0$ is very much reduced. Also indicated ($\times$) is the optimum composition for maximizing the resolution of the most poorly separated band pair. The actual separation based on this optimum mobile-phase composition is shown in Fig. 26 for several nominally similar 25-cm silica columns. The desired resolution ($R_s > 1.0$) is indeed observed for all three columns. This is an important point when retention optimization is applied to complex mixtures; if column-to-column variability in retention is significant, an optimum separation on one column may not be transferable to

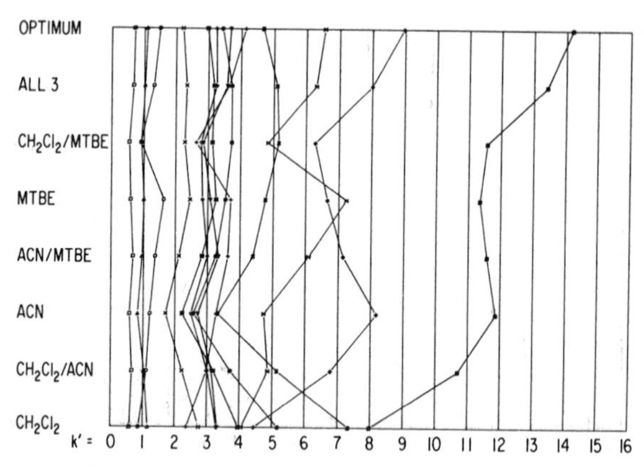

FIG. 24. Retention data for mixture of 13 substituted naphthalenes and 7 mobile phases from Fig. 23 B (hexane present in all 8 mobile phases). See Ref. *20* for details. Reprinted with permission from Glajch *et al.* (*20*).

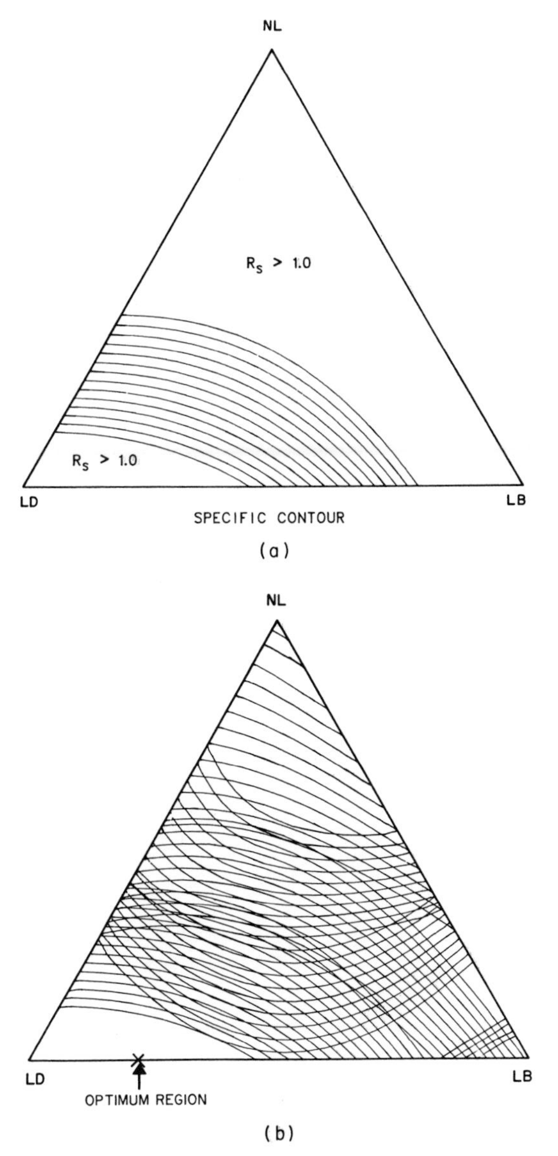

FIG. 25. Overlapping resolution maps for sample of Fig. 24: (a) Band pair 6–8; (b) all band pairs.

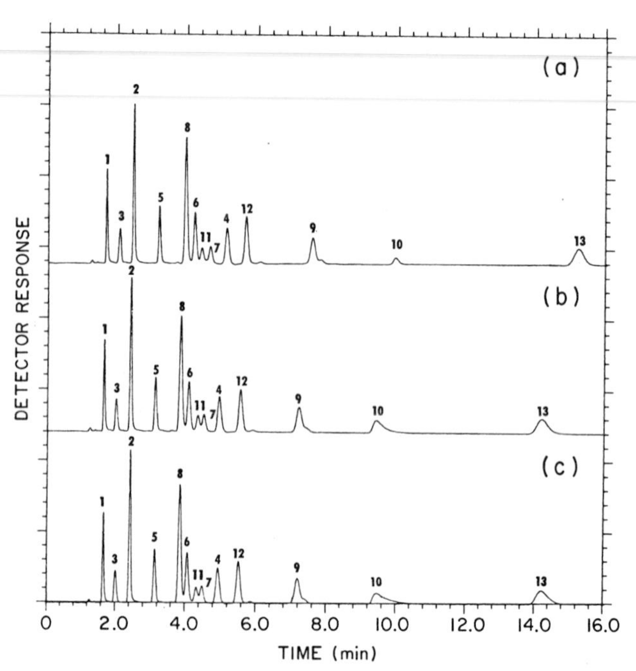

FIG. 26. Separation of mixture of Fig. 24 on three similar columns with optimum mobile-phase composition of Fig. 25b: (a) Coumn 1, used 8 months; (b) column a, new; (c) column 3, new.

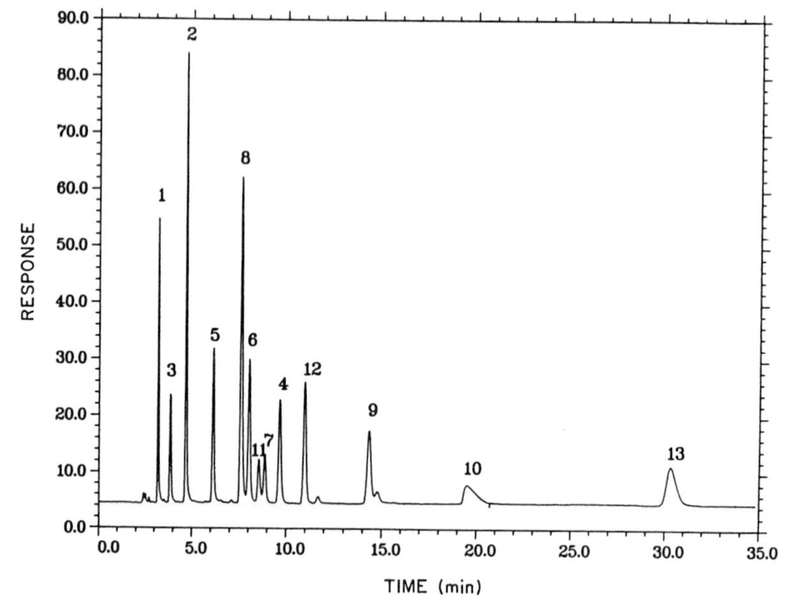

FIG. 27. Separation of Fig. 26 improved by using two columns in series (50 cm as compared to 25 cm in Fig. 25).

another column. This in turn means that all separations via the optimized mobile phase must be carried out on a single column. When columns are reproducible, it is possible to replace a column after its efficiency decreases with time, or the column fails for any reason, without having to go through method development for the new column. Likewise, standard assay procedures can be transferred easily from one lab to the next.

Once retention optimization has been achieved as in Fig. 26, further increase in R_s can usually be achieved by increase in column length (see Ref. *44*). This is shown for the separation of Fig. 26 in Fig. 27, where column length is doubled by connecting two 25-cm columns in series. The time required for separation is now doubled, but the separation of band pair 7–11 is improved, as is the resolution of an impurity on the side of band 9.

IV. CONCLUSIONS

The present article has summarized a comprehensive, yet detailed model of retention in liquid–solid chromatography that allows accurate predictions of solvent strength and solvent selectivity. The model has been tested for several adsorbents: alumina, silica, partially silanized (C_{18}) silica, and amino-alkyl silica. It has likewise been evaluated for a broad range of solvents (about 25 polar solvents plus pentane, hexane, and Freon-113) and mixtures of these solvents containing two to four components. Finally, the model has been tested for numerous solutes of widely varying functionality, including isomeric solute pairs and solutes capable of hydrogen bonding with the solvent. In each of these cases, the variation in retention with change in adsorbent, solvent, or solute can be rationalized in terms of well-known principles of physical organic chemistry and/or thermodynamics.

The model has been further tested in terms of the behavior of solvent molecules: isotherms for binary-solvent mixtures A/B; adsorption energies (ϵ_B values) of the polar solvent B in such mixtures as a function of surface coverage θ_B, etc. Again, good agreement of experimental data with calculated values is observed. An important requirement of the model and related experimental correlations is that the solvent molecule must be treated (thermodynamically) in the same manner as solute molecules. Thus, if *solute* adsorption energies are measured for a molecule X, the behavior of X as a *component* (solvent) of the mobile phase should then be predictable.

The model has also been tested in terms of the known structures of adsorption sites (Fig. 14) in these different adsorbents. Thus, these sites differ in terms of (a) rigid versus nonrigid positioning on the surface, (b) accessibility versus nonaccessibility for covalent interaction, (c) surface concentration (mol/m^2), and (d) chemical functionality. It is found that mobile-phase effects vary from adsorbent to adsorbent, and these differences can be rationalized in terms of structural differences in the adsorption sites.

The present model predicts how solvent selectivity will vary with mobile-phase composition, and this allows the selection of "extreme" solvents for maximum differences in selectivity. This information plus the ability to calculate solvent strength versus composition of the mobile phase then allows development of a general strategy for optimizing retention of any sample, so as to maximize resolution. This four-solvent approach can be further refined by use of computer-assisted procedures, such as the overlapping-resolution-mapping technique.

The basis of the present model is a displacement mechanism of retention, with recognition of solvent and solute localization effects, and consideration of possible hydrogen bonding between solvent and solute molecules in either the adsorbed or nonsorbed phases. Localization is a simple physicochemical phenomenon that is directly related to the energies of adsorption of different functional groups within the solute or solvent molecule. However, localization can be adversely affected by various delocalization processes, and these are more diverse in their dependence on the thermodynamics of retention and the nature of adsorption sites. Three kinds of delocalization have so far been identified and correlated with experimental observation: (a) intramolecular delocalization, where localization of one group k within a molecule affects the adsorption energies of remaining groups $h, i, j, \ldots$; (b) site-competition delocalization, where adjacent nonlocalized molecules can decrease the adsorption energy of a localized molecule; (c) restricted-access delocalization, where only a certain fraction (about 75%) of a monolayer of adsorbed solvent molecules C can adsorb with localization.

The various aspects of displacement and localization are now well understood, and predictions of their effects on retention in LSC can be made with some confidence. Hydrogen bonding between solute and solvent molecules requires further investigation, and it is likely that such studies will contribute to our understanding of hydrogen bonding in solution as well. On the basis of the present model it should prove possible to systematically explore new stationary phase compositions for unique separation potential. However, this subject falls outside the area of mobile-phase effects per se, and will be reserved for another time.

APPENDIX. CALCULATION OF SOLVENT STRENGTH ϵ° FOR MULTICOMPONENT MOBILE PHASES IN LSC

The basis for calculating values of ϵ° as a function of mobile-phase composition is Eq. (11):

$$\epsilon^\circ = \epsilon_A + \log(N_A/\theta_A)/\alpha' n_b \quad \text{(A-1)}$$

where θ_A is given by an extension (*15*) of Eq (39):
for binary solvents

$$\theta_A = [1 + (N_B/N_A)10^{\alpha' n_b(\epsilon_b - \epsilon_a)}]^{-1} \quad \text{(A-2a)}$$

for ternary solvents

$$\theta_A = \{1 + [(N_B/N_A)10^{\alpha' n_b(\epsilon_b - \epsilon_a)}][1 + (N_C/N_B)10^{\alpha' n_b(\epsilon_c - \epsilon_b)}]\}^{-1} \quad \text{(A-2b)}$$

for n solvents

$$\theta_A = \{1 + [(N_B/N_A)10^{\alpha' n_b(\epsilon_b - \epsilon_a)}][1 + (N_C/N_B)10^{\alpha' n_b(\epsilon_c - \epsilon_b)}] \times [1 + \cdots (1 + [1 + (N_k/N_j)10^{\alpha' n_b(\epsilon_k - \epsilon_j)}])]\}^{-1} \quad \text{(A-2c)}$$

Values of ϵ_i and n_b are calculated in advance as described below, and values of N_i will be known initially.

A. Calculations

1. Calculation of n_b values

It is believed intuitively that the best value of n_b for use in Eq. (A-1) is that given by Eq. (11b):

$$n_b = \frac{\theta_B n_b + \theta_C n_c + \cdots}{(1 - \theta_A)} \quad \text{(A-3)}$$

However, this approach to calculating n_b is troublesome, since it requires an iterative approach. A value of n_b must first be assumed in order to calculate values of θ_i for use in Eq. (A-3). It is much more convenient and apparently about as reliable to use a weighted average according to mole fractions N_i:

$$n_b = \frac{N_B n_b + N_C n_c + \cdots}{(1 - N_A)} \quad \text{(A-3a)}$$

Note for the case of a binary-solvent mixture A/B, Eq. (A-3a) [or Eq. (A-3)] gives n_b as the n_b value of solvent B. As an example of Eq. (A-3a), assume a 4-solvent mobile phase with values of n_b and N_i as follows:

	n_b	N_i
A	—	0.2
B	3.0	0.2
C	4.0	0.3
D	4.4	0.3

Application of Eq. (A-3a) to these values of n_b and N_i yields the effective value of n_b for the mixture equal to 3.9.

2. Calculation of ϵ_i Values

In the case of mixtures of nonlocalizing solvents, ϵ_i for each solvent in the mixture is equal to the value of ϵ'' from Table I. For the case of solvent mixtures where either nonpolar solvents ($\epsilon^\circ \approx 0.0$) and/or localizing solvents comprise the mobile-phase mixture, values of ϵ_i for the localizing solvents are obtained from Eqs. (12) and (40). This is an iterative procedure, since intermediate values of θ_i must be calculated. Its application is illustrated in the examples below. For the most general case, where nonlocalizing polar solvents are present in admixture with localizing solvents, the above calculation procedure must be preceded by the calculation of ϵ' values for the various localizing solvents in the mixture, using Eq. (15) and values of $f_l(C)/A_c \equiv f_l(C)/n_b$ from Table I. This procedure is illustrated below in the examples.

B. Examples

1. No Site-Competition Delocalization of Localizing Solvents

In this example we assume that no site-competition delocalization occurs (ϵ° for the A solvent equals zero): a binary mixture of hexane and MTBE with mol fractions of 0.956 and 0.044, respectively, and silica as adsorbent. No n_b averaging is necessary here, since only one modifier is involved, and $n_b = 4.5$ for MTBE (Table I). The values of $\epsilon_a = 0.00$ and $\alpha = 0.57$[22] are used, along with $\epsilon_b' = 1.01$ and $\epsilon_b'' = 0.48$ (Table I). Assuming $\%_{lc} = 0.50$, then values $\epsilon_b = 0.745$, $\theta_a = 0.211$, and $\theta_b = 0.789$ are calculated. From this latter value of θ_b, a new $\%_{lc}$ of 0.454 is calculated from Eq. (40):

$$\%_{lc} = (1 - \theta_b)[1/(1 - 0.94\theta_b) - 14.5\theta_b^9]$$

The procedure is iterated until final values are calculated as follows: $\%_{lc} = 0.482$, $\epsilon_b = 0.736$, $\theta_a = 0.220$, $\theta_b = 0.780$, and $\epsilon_{ab} = 0.249$.

[22] Can be assumed constant for silica whenever moderately polar solutes are involved.

2. Site-Competition-Delocalization of Localizing Solvents

Consider the calculation of ϵ° for a ternary-solvent mixture with hexane ($N_A = 0.7165$), methylene chloride ($N_B = 0.270$), and MTBE ($N_C = 0.0135$), again with silica as adsorbent. The value of α' is 0.57 and $\epsilon_a = 0.00$. The n_b values are $n_b = 4.1$ for CH_2Cl_2 and $n_c = 4.5$ for MTBE. The weighted average n is then calculated to be 4.12. From Table I, $\epsilon_b = 0.30$ for CH_2Cl_2 and the values for MTBE in hexane are $\epsilon_c' = 1.01$ and $\epsilon_c'' = 0.48$. Using Eq. (15), the actual (with site competition delocalization) ϵ_c' for this system can be calculated to be 0.855 for MTBE. A value of $\%_{lc} = 0.50$ is first assumed, which results in an $\epsilon_c = 0.668$. Then $\theta_a = 0.277$, $\theta_b = 0.529$, and $\theta_c = 0.194$. Based on this θ_c value, a new $\%_{lc}$ is determined to be 0.986 by using Eq. (4). This process is iterated until final values are obtained: $\%_{lc} = 0.956$, $\epsilon_c = 0.839$, $\theta_a = 0.194$, $\theta_b = 0.395$, $\theta_c = 0.411$, and $\epsilon^\circ = 0.232$.

See Refs. *14* and *15* for further examples, keeping mind that the complication of site-competition delocalization of localizing solvents was unknown when these articles were written.

LIST OF SYMBOLS

A, B, C, D, . . .	Various pure solvents; A is normally nonpolar and solvents B–D are polar; C and D are normally localizing solvents, and B is usually (but not always) nonlocalizing
A_a, A_n, B_a, B_n, . . .	A molecule (A, B, . . .) in the adsorbed (a) or nonsorbed (n) phase
A_m, A_x, A_y, A_1, A_2	Values of A_s for mobile phase (m), solutes x and y, mobile phases 1 and 2
A_s	Cross-sectional area (1 unit = 0.085 nm²) of a solute molecule [Eq. (8)]
$(A_s)_{expt}$	The apparent value of A_s for a solute molecule, inferred from the application of Eq. (8) to experimental data for the solute; see Eq. (14)
C_1, C_2	Constants [Eq. (31a)] for an LSC system with silica as adsorbent and a given pair of solutes
C_a, C_c	Constants [Eq. (36)] for an LSC system with silica as adsorbent and a given pair of solutes
E_{ia}, E_{ij}	Dimensionless free energy of interaction of some adsorbate *i* with the adsorbed phase (a) or any phase *j*
E_{mn}, E_{xn}	Dimensionless free energy of interaction of a mobile-phase molecule (m) or solute molecule (x) with the nonsorbed phase (n) [Eq. (4)]
E_{ma}, E_{xa}	Dimensionless free energy of interaction of a molecule of mobile phase (m) or solute (x) with the adsorbed phase (a) [Eq. (4)]
$E_{ia}^*, E_{ma}^*, E_{xa}^*$	Values of E_{ia}, E_{ma}, or E_{xa} for an adsorbent of "standard" activity [Eq. (4)], where $\alpha' = 1$
E_{m1}^*, E_{m2}^*	Values of E_{ma}^* for mobile phases 1 and 2 [Eq. (7)]

E°_{xa}	For a localizing solute X, the value of E_{xa} when ϵ° for the mobile phase is zero [Eq. (13)]
$E_{ia}{}^{m}$, $E_{ma}{}^{m}$, $E_{xa}{}^{m}$	An additional contribution to E_{ia}, E_{ma}, or E_{xa} due to interactions of species i, mobile phase M, or solute X with mobile-phase molecules in the adsorbed phase [Eqs. (17c), (17d)]
$f_l(C)$, $f_l(X)$, $f_l(k)$	A localization function (Table II) for localizing molecules C, X, or functional group k [Eqs. (13), (15), (37)]; recognizes site-competition delocalization
$f(\Theta_C)$	A localization function [Eq. (30a)] which recognizes the effect of restricted-access delocalization on localization selectivity; see Fig. 16
k'	Solute capacity factor
k_x, k_y	k' Values of solutes X and Y
k_{xb}, k_{xc}, k_{yb}, k_{yc}	k' Values of solutes X and Y in mobile phases containing B and C, respectively [Eq. (26)]
k_1, k_2	Solute k' value for mobile phases 1 and 2 [Eq. (8)]
K	Adsorption equilibrium constant [Eqs. (2) and (3)]
K_{ba}, K_{cb}	Values of K for equilibrium of solvent mixtures A/B and B/C with adsorbent surface [Eq. (10)]
K_{xb}, K_{xc}, K_{yb}, K_{yc}	Values of K for solutes X and Y; mobile phases contain B or C
K^{cx}, K^{cx*}	Hydrogen-bonding equilibrium constants in mobile phase and adsorbed (*) phase, respectively [Eq. (22)]; for reactions (20)
m	Mobile-phase localization-selectivity function [Eq. (30a)]
m°	Value of m for pure solvent [Eq. (30a); Table I]
m_b, m_c, m_i	Values of m for mobile phase containing B, C, or i [Eq. (31)]
M_a, M_n	Molecule M in adsorbed (a) or nonsorbed (n) phases
n	Stoichiometry of adsorption equilibrium [Eq. (1)]; equal A_s/A_m
n_b	Value of A_m for the mobile phase [see Eqs. (11)–(11b)]
n_B, n_i	Value of n_b for solvents B or i
N_A, N_B, N_C, . . . , N_i	Mole fractions of solvent components A, B, C, . . . , i in mobile phase
P'	Solvent polarity index
Q	Activity coefficient ratio [Eq. (18a)] accounting for effect of solvent–solute interactions on retention (apart from hydrogen bonding)
Q°_k	Adsorption energy of functional group k (localizing) in solute or solvent molecule
R	Gas constant
R_M	Retention factor, equal to log k'
T	Absolute temperature (K)
V_a, V_n	Volumes of adsorbed and nonsorbed phases, per gram of adsorbent
W°	Monolayer uptake (g/g) of solvent B from pure B as mobile phase
X_a, X_n	Molecule of solute X in adsorbed or nonsorbed phases
$(X)_a$, $(X)_n$	Concentration of X in adsorbed or nonsorbed phase
XH	A proton-donor solute
XH_a, XH_n	A molecule of XH in adsorbed or nonsorbed phases
$(XH)_a$, $(XH)_n$	Concentrations of XH in adsorbed and nonsorbed phases
XHC	Complex of solute XH and proton-acceptor solvent C
$(XHC)_a$, $(XHC)_n$	Concentrations of XHC in adsorbed and nonsorbed phases

α Separation factor for compounds X and Y; equal to k_x/k_y
α_b Value of α for mobile phase containing solvent B
α_c Value of α for mobile phase containing solvent C
$(\alpha_c)_{corr}$ Value of α_c for different mobile phases, corrected to a constant $\epsilon°$ value [Eq. (42)]
α' Adsorbent activity function, proportional to adsorbent surface energy; $\alpha' = 1$ for "standard" absorbent
β, γ Parameters defined by Hammers *et al.* (*35, 36*) which measure importance of delocalization due to intramolecular effects (β) or to site competion (γ)
γ_{ij} Activity coefficient for species i in phase j
$\gamma_{ma}, \gamma_{mn}, \gamma_{xa}, \gamma_{xn}$ Activity coefficients for mobile phase molecule (m) or solute (x) in adsorbed (a) or nonsorbed (n) phases (hexane solution as standard state)
Δa_i Increase in A_s $[(A_s)_{expt} - A_s]$ due to localization of a solute group i
Δ_l Solvent–solute localization contribution to retention [Eq. (29)]
Δ_x, Δ_y Solute parameters for solutes X and Y, increasing in value for solutes that are localized to a greater extent [Eq. (30) and Fig. 17]
ΔE_{xm} Dimensionless free energy of adsorption of solute X from mobile phase M [Eq. (3)]
ΔE^*_{xm} Value of ΔE_{xm} for adsorbent of standard activity [Eq. (4)]
$\Delta G°$ Standard free energy for reaction (1)
ΔR_M Increase in R_M as a result of substituting a functional group i into the solute molecule; equal to $\log k_{xi} - \log k_x$, where k_{xi} is for the solute substituted by i and k_x for the solute substituted by $-$H
$\epsilon°$ Solvent strength parameter, equal E^*_{ma}/A_m for solvent M
$\epsilon_b, \epsilon_c, \epsilon_i$ Values of $\epsilon°$ for mobile phases containing solvents B, C, or i [Eq. (28)]
$\epsilon_A, \epsilon_B, \epsilon_C, \ldots$ Values of $\epsilon°$ for pure solvents A, B, C, . . .
ϵ_M Value of $\epsilon°$ for a mixture A/B (mobile phase M)
ϵ_1, ϵ_2 Value of $\epsilon°$ for mobile phases 1 and 2 [Eq. (8)]
ϵ^m Value of $\epsilon°$ corrected for mobile-phase interactions [Eq. (19)]
ϵ', ϵ'' Values of $\epsilon°$ for the pure solvent (localizing) in mobile-phase mixtures; ϵ' is for $\Theta = 0$ and ϵ'' is for $\Theta = 0$ (Table I)
$\epsilon'_B, \epsilon'_C, \epsilon''_B, \epsilon''_C$ Values of ϵ' and ϵ'' for solvents B and C
$(\epsilon'_C)°$ Value of ϵ'_C in a mixture A/C where A is nonpolar ($\epsilon_A \approx 0$)
$\theta_A, \theta_B, \theta_C, \ldots, \theta_i$ Mole fraction of solvents A, B, C, . . . , i in adsorbed phase; Eq. (39)
$\%_{lc}$ A localization function for calculating ϵ_c as a function of Θ_B; Eqs. (12) and (40)

Acknowledgment

Several workers have contributed to clarifying and otherwise improving the present article through their comments and discussion. I am indebted to Drs. Hans Poppe, J. J. Kirkland, M. Caude, and W. E. Hammers. These chromatographers have made a complicated story a bit easier to understand.

REFERENCES

1. L. R. Snyder, "Principles of Adsorption Chromatography." Dekker, New York, 1968.
2. E. Soczewinski, *Anal. Chem.* **41,** 179 (1969).
3. R. P. W. Scott and P. Kucera, *Anal. Chem.* **45,** 749 (1973).
4. R. P. W. Scott and P. Kucera, *J. Chromatogr.* **149,** 93 (1978).
5. R. P. W. Scott and P. Kucera, *J. Chromatogr.* **171,** 37 (1979).
6. R. P. W. Scott, *J. Chromatogr. Sci.* **18,** 297 (1980).
7. M. Jaroniec and A. Patrykiejew, *J. Chem. Soc. Faraday Trans. I* **76,** 2486 (1980).
8. M. Jaroniec and J. Piotrowska, *HRC CC, J. High Resolut. Chromatogr. Chromatogr. Commun.* **3,** 257 (1980).
9. J. K. Rozylo, J. Oscik, B. Oscik-Mendyk, and M. Jaroniec, *HRC CC, J. High Resolut. Chromatogr. Chromatogr. Commun.* **4,** 17 (1981).
10. M. Jaroniec, J. A. Jaroniec, and W. Golkiewicz, *HRC CC, J. High Resolut. Chromatogr. Chromatogr. Commun.* **4,** 89 (1981).
10a. M. Jaroniec and J. Oscik, *HRC CC, J. High Resolut. Chromatogr. Chromatogr. Commun.* **5,** 3 (1982).
11. L. R. Snyder, *Anal. Chem.* **46,** 1384 (1974).
12. L. R. Snyder and H. Poppe, *J. Chromatogr.* **184,** 363 (1980).
13. L. R. Snyder, *J. Chromatogr.* **63,** 15 (1971).
14. L. R. Snyder and J. L. Glajch, *J. Chromatogr.* **214,** 1 (1981).
15. J. L. Glajch and L. R. Snyder, *J. Chromatogr.* **214,** 21 (1981).
16. L. R. Snyder and J. J. Glajch, *J. Chromatogr.* **248,** 165 (1982).
17. L. R. Snyder and T. C. Schunk, *Anal. Chem.* **54,** 1734 (1982).
18. L. R. Snyder, J. L. Glajch, and J. J. Kirkland, *J. Chromatogr.* **218,** 299 (1981).
19. L. R. Snyder, *J. Chromatogr.* (in press).
20. J. L. Glajch, J. J. Kirkland, and L. R. Snyder, *J. Chromatogr.* **239,** 268 (1982).
21. J. L. Glajch and L. R. Snyder, to be submitted.
22. E. Soczewinski and J. Jusiac, *Chromatographia* **14,** 23 (1981).
23. E. Soczewinski and J. Kuczmierczyk, *J. Chromatogr.*, **150,** 53 (1978).
24. S. Hara, Y. Fujii, M. Hirasawa, and S. Miyamoto, *J. Chromatogr.* **149,** 143 (1978).
25. S. Hara, M. Hirasawa, S. Miyamoto, and A. Ohsawa, *J. Chromatogr.* **169,** 117 (1979).
26. S. Hara, A. Ohsawa, and A. Dobashi, *J. Liq. Chromatogr.* **4,** 409 (1981).
27. S. Hara, K. Kunihiro, H. Yamaguchi, and E. Soczewinski, *J. Chromatogr.* **239,** 687 (1982).
28. B. L. Karger, L. R. Snyder, and C. Eon, *Anal. Chem.* **50,** 2126 (1978).
28a. M. Jaroniec and J. Oscik, *J. High. Resolut. Chromatog. Chromatog. Commun.* **5,** 3 (1982).
28b. L. R. Snyder, *Adv. Anal. Chem.* **3,** 288–289 (1964).
29. B. L. Karger, L. R. Snyder, and C. Horvath, "An Introduction to Separation Science." Wiley-Interscience, New York, 1973.
30. L. R. Snyder, *J. Chromatogr.* **8,** 319 (1962).
31. M. C. Hennion, C. Picard, C. Combellas, M. Caude, and R. Rosset, *J. Chromatogr.* **210,** 211 (1981).
32. E. H. Slaats, J. C. Kraak, W. J. T. Brugman, and H. Poppe, *J. Chromatogr.* **149,** 255 (1978).
33. J. S. Perry, *J. Chromatogr.* **165,** 117 (1979).
34. L. R. Snyder, *J. Chromatogr.* **6,** 22 (1961); **8,** 178 (1962).
35. W. E. Hammers, R. H. A. M. Janssen, A. G. Baars, and C. L. de Ligny, *J. Chromatogr.* **167,** 273 (1978).

36. W. E. Hammers, M. C. Spanjer, and C. L. de Ligny, *J. Chromatogr.* **174,** 291 (1979).
37. M. C. Hennion, C. Picard, C. Combellas, M. Caude, and R. Rosset, *J. Chromatogr.* **210,** 211 (1981).
38. M. D. Palamareva, B. J. Kurtev, M. P. Mladenova, and B. M. Blagoev, *J. Chromatogr.* **235,** 299 (1982).
39. L. R. Snyder, *J. Chromatogr.* **245,** 165 (1982).
40. L. R. Snyder, *J. Chromatogr.* **92,** 223 (1974).
41. L. R. Snyder, *J. Chromatogr. Sci.* **16,** 223 (1978).
42. S. Hermanek, V. Schwartz, and Z. Cekan, *Collec. Czech. Chem. Commun.* **28,** 2031 (1963).
43. V. Prey, H. Berbalk, and H. Krünes, *Mikrochim. Ichnoanal. Acta* p. 333 (1964).
44. L. R. Snyder and J. J. Kirkland, "Introduction to Modern Liquid Chromatography," 2nd ed., Chapter 2. Wiley-Interscience, New York, 1979.
45. J. L. Glajch, J. J. Kirkland, and W. G. Schindel, *Anal. Chem.* **54,** 1276 (1982).
46. J. L. Glajch, J. J. Kirkland, K. M. Squire, and J. M. Minor, *J. Chromatogr.* **199,** 57 (1980).

INDEX

R

S

T

U–Z